The Seaward Margin of Belize Barrier and Atoll Reefs

Special Publication Number 3 of the International Association of Sedimentologists
by Noel P. James and Robert N. Ginsburg
published by Blackwell Scientific Publications
Oxford London Edinburgh Boston Melbourne

The Seaward Margin of
Belize Barrier and Atoll Reefs

To T. W. Edgeworth David, leader of the Second and Third Expeditions to Funafuti Atoll, Ellice Islands; 1897 and 1898. Pioneer researcher on the seaward margin of coral reefs (photograph courtesy Harry Ladd).

The Seaward Margin of Belize Barrier and Atoll Reefs

Morphology, Sedimentology, Organism Distribution and Late Quaternary History

NOEL P. JAMES & ROBERT N. GINSBURG

Comparative Sedimentology Laboratory
Rosenstiel School of Marine and Atmospheric Science
University of Miami, Fisher Island Station
Miami Beach, Florida 33139, USA
Noel P. James is now at: Department of Geology,
Memorial University of Newfoundland, St John's,
Newfoundland A1B 3X5 Canada

SPECIAL PUBLICATION NUMBER 3 OF THE
INTERNATIONAL ASSOCIATION OF SEDIMENTOLOGISTS
PUBLISHED BY BLACKWELL SCIENTIFIC PUBLICATIONS
OXFORD LONDON EDINBURGH MELBOURNE

B

PRIS

Published by Blackwell Scientific Publications
Osney Mead, Oxford, OX2 0ES
8 John Street, London, WC1N 2ES
9 Forrest Road, Edinburgh, EH1 2QH
214 Berkeley Street, Carlton, Victoria 3053, Australia

First published 1979

James, Noel P
The seaward margin of Belize barrier and
atoll reefs. – (International Association of
Sedimentologists. Special publications; no. 3).
I. Coral reefs and islands – Belize
I. Title II. Ginsburg, Robert Nathan
III. Series
551.4′2 QE565

ISSN 0141-3600
ISBN 0 632 00523 8

Distributed in the U.S.A. by
Halsted Press, a division of
John Wiley & Sons, Inc., New York

Printed and bound in Great Britain by
Burgess & Son (Abingdon) Ltd
Station Road, Abingdon, Oxfordshire

Contents

Preface

Fossil reef complexes and carbonate platforms contain more than a third of the world's reserves of petroleum, an important share of certain metallic ores, and an unusually sensitive and legible record of earth history and the evolution of life. Models for interpreting fossil platforms have come from studies of the morphology, sediments and diagenesis of modern platforms and continental shelves. This comparative approach has had its more notable success in shallow-water deposits. For example, from studies of Holocene tidal flat sediments it is now possible to recognize in carbonates as old as Precambrian the precise limits of tidal zones (Hoffman, 1973; Klein, 1971). The same studies have demonstrated that the formation of dolomite may accompany deposition (Pray & Murray, 1965), and they have shown how knowledge of tidal sedimentation provides the basis for understanding and predicting the occurrence of potential reservoirs for oil and gas (Lucia, 1972).

Research on ancient buried platforms and reef complexes guided by first-generation models from present-day examples has led to the realization that the marginal zone—the complete transition from platform interior or reef lagoon across the rim and down the slope to deeper water—is the most critical element of the entire complex. This zone is a major faunal, stratigraphic, and sedimentological discontinuity; it is the locus of reefs, bioherms and buildups of lime sand that may influence deposition on the platform interior; and it is the preferred site for reservoirs of oil and gas and accumulations of metallic ores.

Valuable as studies of fossil platform margins are, and certainly more are needed, they can only reveal the end product of deposition, modified by successive stages of diagnesis; they can show what has happened where, but not how or when. Yet if models of the marginal zone are to be developed, knowledge of the how and when is essential.

The success of the comparative approach in other parts of the platform and in siliciclastic sediments mandates that a necessary step in the development of models is thorough study of present-day examples. Such studies are not now available and it is to fill that need that this monograph is directed. Its principal purpose is to present results of our study of the fore-reef zone of the barrier and atoll reefs of Belize, a study based primarily on direct observations and collections made from a research submersible combined with seismic profiling and examination of piston cores from the adjacent basinal deposits.

Observations from the submersible have provided exact information as to the different kinds and depth ranges of carbonate producing organisms, the morphology of the fore-reef zone and the distribution of sediments. Examination of precisely located samples of reef-wall and fore-reef limestones has revealed the remarkable, reiterated internal deposition, cementation and boring that have produced several meters of well cemented Holocene accretion. Study of piston cores and of sediments taken using the submersible has allowed us to define the surprisingly narrow zone of reef-influenced sedimentation along the basin margin and study of seismic profiles gives indications as to the third dimension of this transition.

Fortunately these new findings on the fore-reef can be set in their proper perspective as we are able to draw on the comprehensive reports of Edward Purdy and his students on the modern environments and Holocene sediments of the shallow-water, barrier-reef platform and lagoon, on our own earlier research on the shallow reefs, and on the work of others on the regional geology of the Belize region and the offshore areas. In this way we are able to put our study of the Belize fore-reef in its regional historical and environmental setting. Our results provide an example of deposition and early diagenesis that can guide research on and interpretation of comparable fossil platform margins and an example that can help to develop the needed models of this depositional realm.

This ongoing research on the Belize reefs was supported by National Science Foundation Grant Number GA-29302 and funds for the operation of the submersible NEKTON were provided by contract number 3-35218 from the Manned Undersea Science and Technology Program of the National Ocean and Atmospheric Administration. Preparation of the final manuscript was partially supported by National Research Council of Canada Grant Number A-9159.

This research was carried out with the permission of the Government of Belize and we are particularly grateful to the Minister of Trade and Industry, Mr A. A. Hunter, for granting permission and assisting us in securing supplies and equipment.

The study itself could not have been carried out or completed without the kind and generous assistance of many people. Robert Dill gave us constant advice and encouragement in planning the submersible operations. The skilful piloting of the submersible and acute observations by Richard Slater and Doug Privett were a major factor in the success of our field program. Captain Tom Crawford and crew of R/V SEAMARK were invaluable because of their skilful and willing help to us at all hours. The advice, assistance, experience and constant questioning of the group of colleagues who joined us in the field, Michael Brady, Patrick Colin, Paul Enos, W. Jerry Koch, Willard Hartman, Lynton Land, Judy Lang, Donald Marszalek, Eric Mountjoy and Jack Wray, helped us immeasurably. The success of our sampling program was largely due to the advice on and skilful handling of explosives by Lt A. Y. Bryson, United States Navy, and the dexterity of Richard Davies with the manual claw at great depths.

The geophysical seismic lines were run under the supervision of Paul Enos who subsequently interpreted the records and has contributed to this volume by writing the major portion of Chapter Two. The expertise of Judy Lang and Willard Hartman in identifying and cataloguing most of the deep water fauna has proved invaluable. Michael Brady and Jerry Koch did most of the coring and preliminary analysis of the samples while Donald Marszalek analysed the silt-sized sediment fractions.

Philip Choquette, Michael Lloyd and Lynton Land kindly analysed selected samples for trace elements and stable isotopes. Gerry Stipp supervised the age dating of our samples at the University of Miami Radiocarbon Laboratory. Wolfgang Schlager generously gave us his compilation of platform margin topography for inclusion in Chapter 7.

The samples were slabbed and polished by Cecil Daniels. The 446 thin sections used in this study were prepared by Marathon Oil Company by K. Bolus and stained using the techniques developed by Philip Choquette and Fred Trusell. The development and printing of numerous plates and diagrams was undertaken enthusiastically by Wilfred Marsh. Some of the figures were drafted at Marathon Oil Company and by Eileen Stevens and Robert Hiscock at Memorial University. The cover was designed and illustrated by Stewart Moss ASC, Educational Television, Memorial University. Typing of the numerous tape transcriptions and manuscript drafts was tirelessly tackled by Lois Keith and Glenys Woodland. The final manuscript editing was done by Judith James.

The manuscript was reviewed by three colleagues. We thank Terry Scoffin for his close attention to numerous details. Many of the biological aspects were clarified by Judith Lang's searching critique. We owe special thanks to Lynton Land for his incisive and astringent questioning of observations and interpretations.

February, 1979

Noel P. James
St John's, Newfoundland
Robert N. Ginsburg
Miami, Florida

Chapter 1

The geological setting of Belize reefs

INTRODUCTION

In the context of modern carbonate platforms and shelves, the Belize region, like Australia's Great Barrier Reef, is a rimmed shelf (Ginsburg & James, 1976). The Belize rim is for most of its length a barrier reef that extends some 250 km from the Yucatan Peninsula to the Gulf of Honduras (Figs 1-1, 1-2). Seaward of the barrier reef, there are three isolated platforms: Glovers Reef, an atoll; Lighthouse Reef with a fringing reef and very shallow lagoon; and Turneffe Islands, mostly islands and mangrove swamps. Between the barrier reef and the mainland lies a lagoon that is 20–40 km wide and which deepens from a few metres in the north to 50 m towards the southern, open end (Fig. 1-2). There are no reefs in the shallow northern lagoon, but in the deeper and wider southern part, patch reefs of various sizes and larger platform atolls or faros are so numerous that navigation is hazardous.

There are two topographic modes seaward of the barrier reef and around the offshore platforms: one in which the steep fore-reef slope extends without major interruption to 1000 m or more, and a second in which the fore-reef slope flattens out at a depth of between 300 and 400 m and passes into a flat or gently sloping bottom. The first type is seen in the southernmost part of the barrier reef where the deep Cayman Trough is both parallel and adjacent to the barrier; on the eastern side of Glovers Reef and Turneffe Islands; and on both sides of Lighthouse Reef (Fig. 1-2). Areas where the fore-reef slope flattens markedly in depths of 300–400 m occur between Glovers Reef and the barrier reef along the barrier north of Turneffe Islands (Fig. 1-2).

These variations in offshore bathymetry and the variations between barrier and atoll reef types were the basis for our choice of study area. Within the southern half of the reef complex we were able to choose locations where the fore-reef extends without major interruption to 1000 m or more both on the barrier reef and on the eastern side of Glovers. As examples of fore-reef zones that terminate near 400 m and pass into flat basin floors we selected sites on the western side of Glovers and on the barrier reef west of Glovers.

At each of seven sites we used the submersible to examine and sample the fore-reef zone to depths of 300 m. For some of these transects we made seismic profiles extending to the offshore troughs; for two of the most thoroughly studied transects we collected gravity and piston cores of the distal fore-reef and adjacent basin.

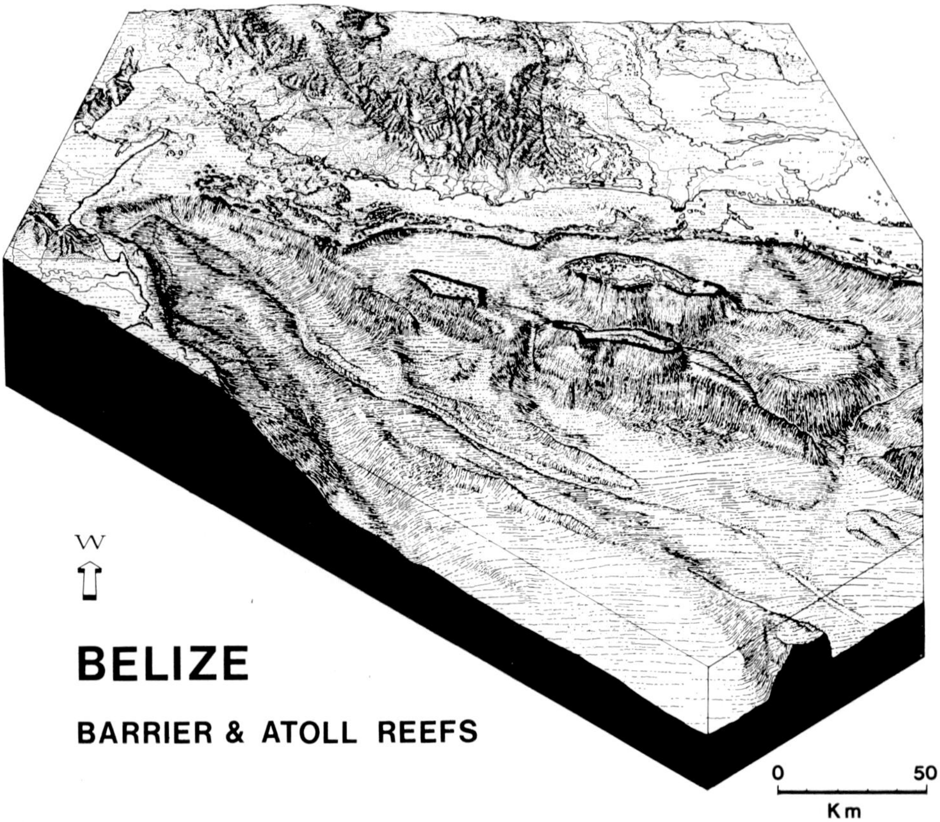

Fig. 1-1. An isometric diagram of the Belize continental margin, looking west, across the Cayman Trough, with the three offshore reef complexes and atolls in front of the long narrow continental shelf and Maya Mountains on the mainland at the upper left (*after* Dillon & Vedder, 1973).

To set the stage for a detailed account of our findings in the fore-reef zone we have, in the following sections of this introductory chapter, outlined the regional aspects of the Belize complex, focusing on the nature of the modern shallow-water reef complex, the record of Holocene sedimentation and the style of the underlying geological and structural framework.

CLIMATE AND WATER CHARACTERISTICS

Climate

The climate of Belize is subtropical (Wright *et al.*, 1959); summer air temperatures average 27°C, and winter air temperatures are a little cooler averaging 24°C. The amount of rainfall reflects the mainland topography; the flat northern part of the country receives on the average less than 25 cm year^{-1} while the mountainous southern region receives up to 70 cm year^{-1}. Prevailing winds blow from the east and average 15 km h^{-1}.

There are two levels of interruption to the normal trade wind regime that affect reef biota and sedimentation in shallow water. Incursions of masses of cold air from

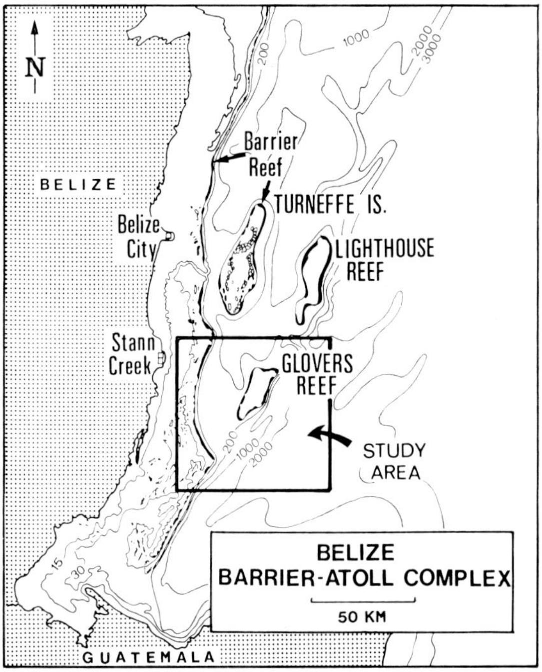

Fig. 1-2. A chart of the Belize continental margin with the area of this study outlined (depth contours in metres).

continental North America from October through January bring periods of a few days of strong northerly winds and heavy rains. The effects of these 'northers' on reefs and shallow-water sediments have not been properly evaluated, but judging from observations in South Florida, these winter storms may be the principal cause of accumulations of reef rubble and movement of rhodolites (nodules of crustose coralline algae) on the reef flat. More devastating than the winter storms are hurricanes that move across the Belize–Yucatan area in summer or early fall. Between 1931 and 1961 the Belize reefs were struck by a major hurricane on the average of once every 6 years (Stoddart, 1963). Surveys of individual reefs before and after Hurricane Hattie of 1961 have shown that it destroyed large areas of branched corals in shallow water and created and destroyed sand cays on the barrier platform (Stoddart, 1963, 1969).

Water characteristics

Continental shelf

The general patterns of salinity in surface waters on the continental shelf have been recorded by Pusey (1964) and reported by Wantland & Pusey (1971), Purdy (1974b) and Purdy, Pusey & Wantland (1975). In summarizing the measurements, Purdy (1974b) points out that two trends are apparent; a decrease north into Chetumal Bay and south into the Gulf of Honduras because of fresh water runoff from land; slightly higher salinity in the north lagoon because of less rainfall than in the south. In the southern lagoon wedges of fresh water occasionally reach out to the barrier reef (Purdy *et al.,* 1975).

Open ocean

Surface circulation in the western Caribbean is dominated by the Caribbean current that approaches the Belize continental margin from the east and is deflected northward to pass between the Yucatan platform and Cuba (Fig. 1-3).

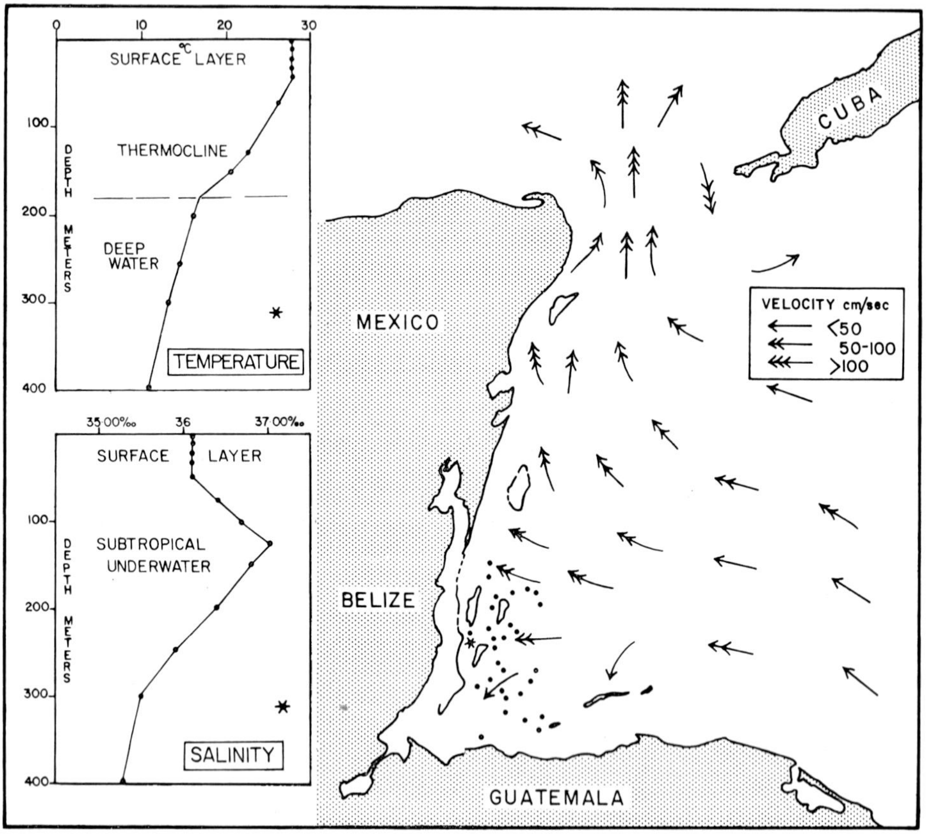

Fig. 1-3. A map illustrating the main surface circulation pattern and current velocities of the western Caribbean Sea (*from* Wust, 1964). The two insets at the left are water temperature and salinity profiles for a station (star) located between the barrier reef and Glovers Reef (data from R. Molinari, personal communication).

The temperature and salinity variations to a depth of 500 m were obtained from a series of thirty stations taken off Belize from R/V RESEARCHER (R. Molinari, personal communication) and R/V DISCOVERER (R. Starr, personal communication) in July 1971 and other data recorded in the National Oceanographc Data Center X–BT listings between October, 1966 and May, 1967. All stations record a well stratified water mass (Fig. 1-3): a well mixed surface layer of isothermal water varying between 27 and 28°C extends to a depth of 50 m; the thermocline, a steep decrease in temperature to about 18°C occurs between 50 and 200 m (the base of the thermocline appears to rise slightly in an easterly direction away from the shelf); Caribbean Deep Water with temperatures between 18 and 10°C is present below a depth of about 200 m.

Surface waters have a similar salinity throughout, ranging between 35·7‰ and 36·1‰ and appear well mixed and isohaline to a depth of 30–50 m (Fig. 1-3). Below this depth is a core of relatively high salinity water with a salinity maximum of 37·0‰ occurring between 100 and 150 m (Fig. 1-3). This subtropical intermediate salinity maximum (Wust, 1964), formed at the surface of the Central Atlantic at 20–25°N and 10–15°S where evaporation is high and precipitation is low, is transported westward by the subtropical undercurrent through the Antilles into the Caribbean.

THE MODERN BARRIER REEF TRACT

Bathymetry

The northern part of the continental shelf is bordered by a low karsted surface of flat-lying Cenozoic and Cretaceous carbonates with only a few small rivers draining onto the shelf. The inner shelf is likewise a flat shallow depression that is rarely deeper than 8 m.

The southern part of the shelf is flanked by the Maya Mountains, which are surrounded by a coastal plain of terrigenous clastic sediments through which many rivers flow onto the shelf. The bathymetry of the inner shelf reflects this mainland topography, deepening southward from a depth of 20–25 m off Belize City to over 200 m in the Gulf of Honduras.

The entire margin of the shelf is a raised barrier that is 3–10 km wide and rarely deeper than 3 m. The precipitous seaward margin of this barrier is crowned by an almost continuous coral reef while the open platform is dotted with small sand cays. The barrier narrows to less than 100 m near its southern end and terminates as a hook-shaped structure in the Gulf of Honduras (Stoddart, 1963, 1969).

Reefs

Despite the stunning array of reefs on the continental shelf and on the offshore atolls they have only been described in a general way (Wantland & Pusey, 1971; Miller & Macintyre, 1977); the only detailed studies are those of James *et al.* (1976) and Halley *et al.* (1977) on the shallow barrier reefs and Wallace & Schafersmann (1977) on the patch reefs of Glovers Atoll.

Lagoon reefs

The innumerable reefs in the south lagoon are of two main types: (1) patch reefs and (2) elongate to rhomboid platform atolls or faroes.

Patch reefs range from small clumps of coral on the barrier platform to zoned reefs up to 80 m across and rising some 20 m from the lagoon floor. Most exhibit a zonation of corals: a seaward-facing steep front slope covered by columnar *Montastraea annularis*; a shallow seaward-facing crescent of *Millepora* sp. and robust, branching *Acropora palmata*; a flat top of coral rubble commonly leading landward to a meadow of the marine grass *Thalassia testudinium*; a gentle leeward slope populated by thickets of delicate, branching corals such as *Acropora cervicornis* and *Porites porites*. The entire patch reef is generally surrounded by an apron of *Halimeda*-rich sand and coral rubble.

Platform atolls have steep slopes, narrow, flat tops and commonly exhibit the same coral zonation as the patch reefs. The centres of these atolls are commonly as deep as the surrounding lagoon floor.

Barrier reef

The shallow barrier reef, as seen from the air, is composed of a rubble-strewn pavement or reef flat fronted by a narrow zone, tens of metres wide (the reef crest), that in turn extends seaward another 100 m or so as a series of spurs and grooves (Fig. 1-4). The spur and groove is the shallow part of the reef front, that slope which extends to a depth of about 65 m. Below this the wall extends to a depth of about 120 m where it is buried by the fore-reef (see Fig. 3-2). The shallow reef flat is composed of coral rubble transported landward from the reef where it is often cemented to form a pavement or swept together to form small cays.

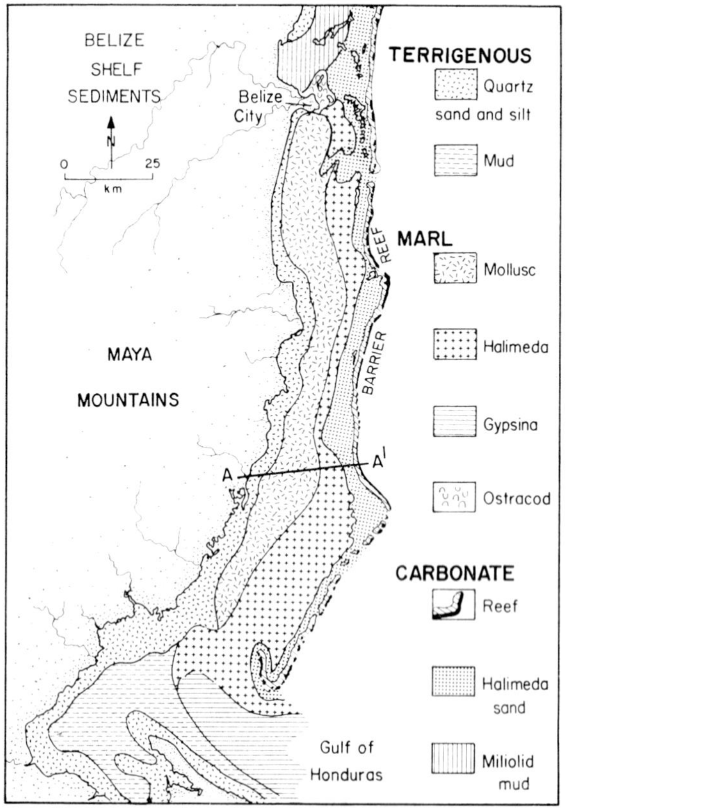

Fig. 1-4. The areal distribution of surficial sediments on the southern part of the Belize shelf (*after* Wantland & Pusey, 1971; Purdy, 1974b).

Underwater, the reef crest and spurs are seen as a lush growth of branching and foliose corals (particularly *Acropora palmata, Porites porites* and *Agaricia agaricites*) and *Millepora* sp. Much of the coral is encrusted with zooanthids and coralline algae. The grooves between spurs widen and deepen seaward and are floored with coralline algae-encrusted coral rubble (particularly *A. palmata*) which spills out and around the front of the spurs onto a narrow terrace-like feature at a depth of 4–5 m. This terrace leads to a second series of spurs and grooves that extend seaward into deeper water. The upper parts of these spurs, with less relief than the shallower set, have a diverse coral fauna and particularly common are the branching species *Acropora cervicornis*, foliose species *Agaricia agaricites* and massive species *Montastraea annularis, Montastrea cavernosa* and *Portes astreoides*. The wide grooves between are floored with coral sticks (mostly *A. cervicornis*) and clean *Halimeda*-rich sands. This description could equally well describe the well studied fringing reef off Jamaica (Goreau, 1659).

Surface sediments

The surface sediments of the continental shelf have recently been documented by Purdy (1974a, b) and Purdy *et al.* (1975) with more detailed studies of specific areas reported by Pusey *et al.* (1975). This work also contains studies by Robertson on the benthic molluscs, by Teeter on the ostracods and by Wantland on the benthonic foraminifers found in surface sediments. Sediments related to the barrier reef have been studied by James *et al.* (1976).

The shallow part of the shelf, north of Belize City, separated from the open ocean by the barrier reef, scattered islands, or the southern extension of the Yucatan Peninsula is floored by pelleted muds containing abundant foraminiferal tests (Fig. 1-4). Terrigenous sediments are restricted to the nearshore zone.

Surface sediments in the southern lagoon, forming the shallow part of the shelf adjacent to the area under discussion in this monograph have been subdivded by Purdy *et al.* (1975) into three separate 'facies' (Fig. 1-4): (1) a nearshore terrigenous facies, (2) an axial marl facies, and (3) a marginal carbonate facies, the main attributes of which are summarized below.

The shoreline and shallow nearshore zone are mantled by quartzose sands, mainly reworked Pleistocene deposits, with local accumulations of terrigenous mud in deltas and a background of reworked and indigenous mollusc debris.

The axis of the lagoon is floored by marl, fine-grained sediment that is transitional in composition between carbonate and terrigenous end members (Purdy *et al.,* 1975). Two trends are apparent in the compositional variation of this marl: (1) a west–east increase in carbonate content from about 30% near the mainland to almost 90% adjacent to the barrier platform; (2) a north–south gradation in the kinds of large skeletal particles: molluscs and foraminifers in the north; plates of the green alga *Halimeda* in the centre; fragments of the encrusting foraminifer *Gypsina* in the Gulf of Honduras at the southern end. The terrigenous, mud-sized fraction is composed largely of clay minerals, specifically kaolinite, illite and montmorillonite, with the latter becoming relatively more abundant away from the mainland. The fine-grained carbonate fraction is derived from the breakdown of larger carbonate skeletons (Matthews, 1965) and fallout of calcareous nannoplankton (Scholle & Kling, 1972).

On the barrier reef platform the composition and texture of the sediments are very distinctive and reflect closeness to the reef. Near the reef crest the shallow sea floor is a pavement of algal-encrusted coral-rubble strewn with cobble to boulder-size pieces of coral debris. Leeward, this flat grades into a sediment apron of lime sand, which in many places is transgressing shelfward over adjacent barrier platform sands. The sediments of the barrier reef, reef flat and sediment apron are composed predominantly of coral, coralline algae and *Halimeda* particles (Coralgal sands). Even though *Halimeda* particles are more abundant than either coral or coralline algae particles in this sediment Purdy *et al.* (1975) prefer to call these 'coralgal' sands because of their very restricted occurrence and the fact that corals and coralline algae particles together exceed the number of *Halimeda* particles. Where passes occur through the barrier reef the sediment apron is very narrow, probably because return flow (seaward) of water breaking on the reef transports sediment into the reef passes and basinward through the reef. Sediments that floor most of the barrier platform are predominantly *Halimeda* sands, with only minor amounts of coral and coralline algae.

The hundreds of isolated patch reefs and platform atolls that rise to the surface from the floor of the lagoon are surrounded by halos of sand-size carbonate that grade into terrigenous-rich mud away from the reefs.

GLOVERS REEF

Glovers Reef is an oval-shaped atoll located 15 km east of the barrier reef, thus protecting part of the shelf somewhat from open ocean waves and swell. The structure is 28 km long and 10 km wide. An almost continuous barrier reef encloses a lagoon that ranges from 6 to 18 m deep and contains over 600 patch reefs. The geological aspects of this atoll have not been studied in nearly as much detail as the barrier reef; the literature is restricted to descriptions of the morphology by Stoddart (1962b), the shallow marginal barrier reef by James *et al.* (1976), and the patch reefs in the lagoon by Wallace & Schafersmann (1977).

The marginal zone of Glovers Reef is similar to that along the barrier reef, except that many of the shallow spurs are essentially bare of living coral and are encrusted with coralline algae. This is probably because this zone, facing the open Caribbean, receives the full force of the numerous hurricanes that sweep through the area (Stoddart, 1963). The marginal reef on the leeward side is narrower; the spurs and grooves are not as well developed; the amount of rock covered with coralline algae is less and *Acropora palmata* is not as common as on the barrier reef.

All of the reefs in the lagoon are patch reefs, similar to those described in the south lagoon of the barrier reef. Detailed study of these patch reefs has revealed a variation in the zonation of corals depending upon the location of the reef in the lagoon (Wallace & Schafersmann, 1977).

Modern sediments associated with the patch reefs are poorly sorted coarse-grained carbonates, composed primarily of *Halimeda,* coral, coralline algae, mollusc and other miscellaneous skeletal particles. The lagoon itself is floored by muddy, very fine-grained carbonate sand, with the sand fraction rich in *Halimeda*, mollusc and foraminifer grains.

THE HOLOCENE SEDIMENTARY RECORD

Seismic profiles and drilling by Purdy (1974b) and Halley *et al.* (1977) reveal that the barrier platform and many of the lagoon reefs are Holocene accretions over pre-Holocene highs (Fig. 1-5). The amount of Holocene accretion on these highs ranges 8 to 20 m and averages 12 m.

Gradual flooding of the lagoon during the Holocene sea level rise began at the southern end, and the sea did not begin to flood the northern shelf until it rose within 10 m of the present level. Throughout most of this time the southern lagoon may have been a large estuary, bounded on the seaward side by the elevated ridge of limestone

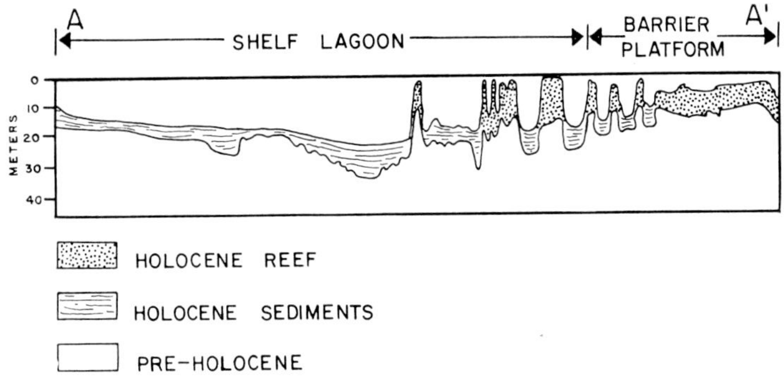

Fig. 1-5. A sketch of a reflection seismic profile across the Belize continental shelf illustrating the pre-Holocene carbonate surface, Holocene open shelf sediments and the Holocene shelf reefs, which appear to be located on pre-existing highs (*after* Purdy, 1974b). For the location of this profile see Fig. 1-4.

that was the Pleistocene barrier, and receiving, for the most part, terrigenous mud from the mainland. Reefs developed and the effect of the barrier was lessened only when sea level rose above the tops of the karst highs some 20 m or so below present sea level about 8000 years B.P. (Purdy, 1974b). Only at this time would production of carbonate sediments have increased dramatically. The total thickness of sediment in the southern lagoon is in the order of 3–5 m.

During almost all of sea level rise the northern shelf was a coastal plain floored by terrigenous clastics. Today the underlying terrigenous sediments are being mixed with contemporary muds by burrowing (Purdy, 1974b).

UNDERLYING GEOLOGICAL AND STRUCTURAL FRAMEWORK

Mainland Belize

Belize straddles two of the major morphotectonic units of Central America: in the south the Sierras of Northern Central America; in the north the Yucatan Platform (Fig. 1-6). The Sierras extend into Belize as the Maya Mountains, a rugged area with peaks up to 1100 m high and mantled by tropical rain forest. These mountains are composed of unmetamorphosed to slightly metamorphosed late Palaeozoic sedimentary rocks and granitic intrusions (Fig. 1-7). The northern, western and southern flanks of the mountains are onlapped by gently dipping to flat-lying Cretaceous limestones. These limestones in turn disappear beneath a yoke of coastal plain sediments in the east, dip under Tertiary terrigenous clastic rocks in the south and are buried by Tertiary carbonates in the north. The northern part of Belize, underlain by the flat-lying carbonates of Cenozoic age, is a low doline karst surface with many shallow dish-like depressions, swamps and sluggish rivers.

The unmetamorphosed to slightly metamorphosed Pennsylvanian and early Permian clastic sedimentary rocks that form the core of the Maya Mounts and extend eastward as part of the Sierras into central Guatemala are believed to form the basement that underlies both northern Guatemala and Belize (Dengo & Bohnenberger, 1969). The geological evolution of the area is characterized by two major orogenic events, one in late Permian to Triassic time and the other in late Cretaceous time. The Permian to late Triassic event that uplifted and deformed the late Palaeozoic

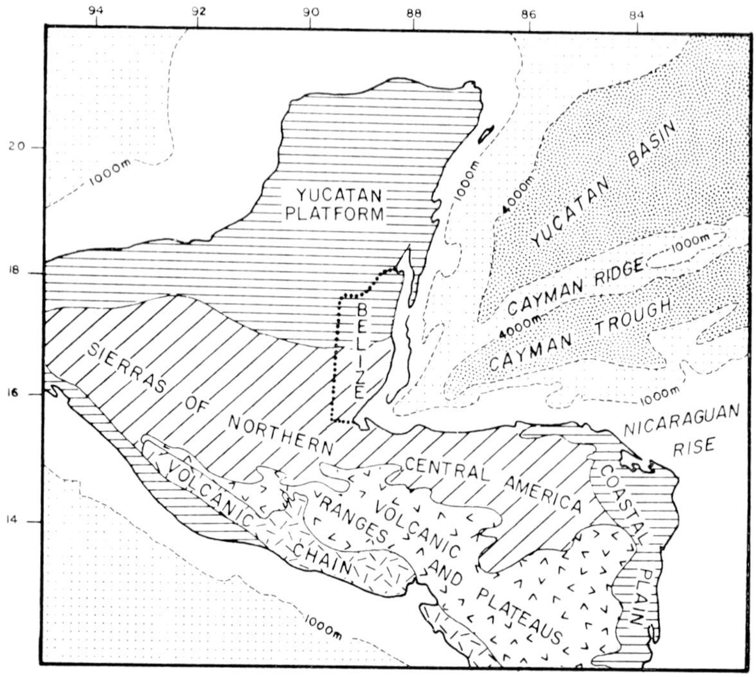

Fig. 1-6. A sketch map of southern Mexico and northern Central America depicting the main morphotectonic elements and submarine topographic units.

sediments was accompanied by intrusion of granitic stocks (Dengo, 1969) and followed by block faulting. Continental redbeds, considered to be late Jurassic and early Cretaceous in age (Dengo & Bohnenberger, 1969) were subsequently deposited in the resulting grabens. During most of the ensuing Cretaceous the whole area, possibly including the Maya Mountains (Purdy, 1974b) subsided and was the site of widespread platform carbonate deposition with accompanying evaporites in northern Belize and Mexico. The presence of overlying normal marine carbonates throughout the region suggests return to normal marine conditions during the late Cretaceous (Viniegra, 1971).

The late Cretaceous early Tertiary event was centred in the Sierras where sediments were folded and thrust northward and again granite plutons emplaced (Dengo & Bohnenberger, 1969). This event is widespread in the Caribbean (Dillon & Vedder, 1973) as evidenced by mobilization of serpentine around the Caribbean margin, intrusion of granodiorites in the Nicaraguan rise, and orogeny, volcanism and basin development in Cuba. Following orogeny, marine redbeds, siltstones and carbonates were deposited throughout Guatemala, Belize and Yucatan.

Early Palaeocene and Eocene deposition comprises a series of sediments in a trough across Guatemala and Southern Belize with reef associated carbonates deposited on the Yucatan Platform. Late deposition comprises a series of localized lacustrine sediments in western Guatemala and southeastern Mexico grading eastward to deltaic and carbonate sediments in restricted embayments of southern Belize and eastern Guatemala.

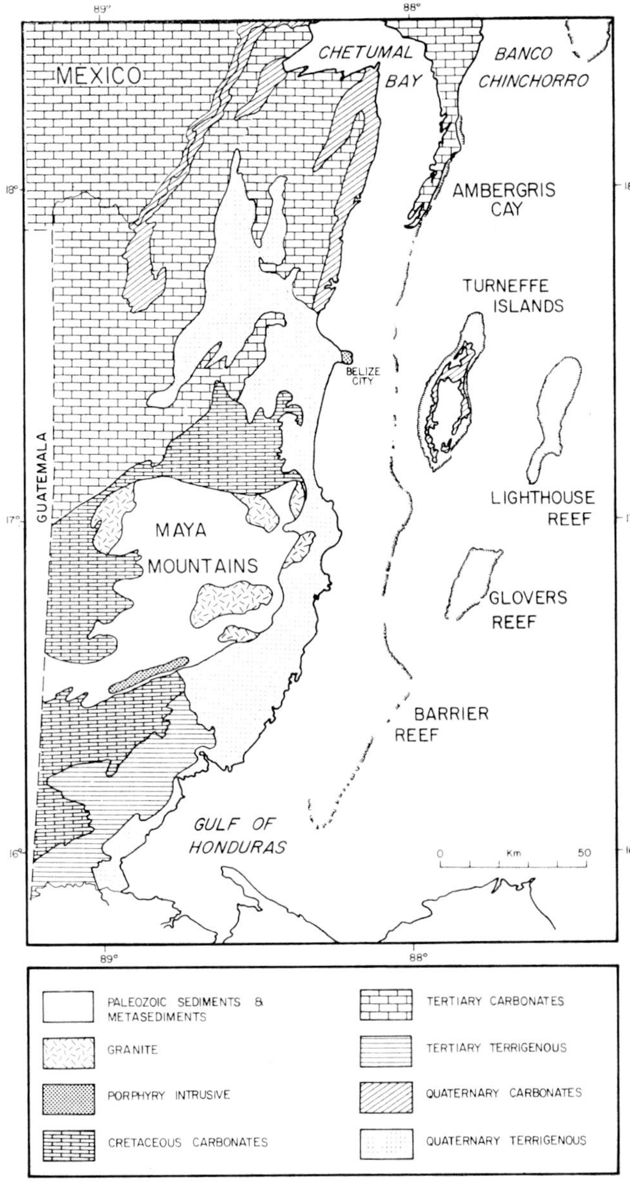

Fig. 1-7. A geological map of mainland Belize (*after* Wantland & Pusey, 1971; Purdy, 1974b).

Continental shelf and adjacent deep sea

The continental margin is characterized by a series of discontinuous ridges that trend northeast (Stoddart, 1963; Baie, 1970). Dillon & Vedder (1973) have identified five of these ridges, three of which are shallow and control the morphology of the shelf and two which are in deeper water and partially buried (Fig. 1-8). Although these ridges diverge northward they are generally parallel to the major rivers in northern Belize and thought to be fault controlled.

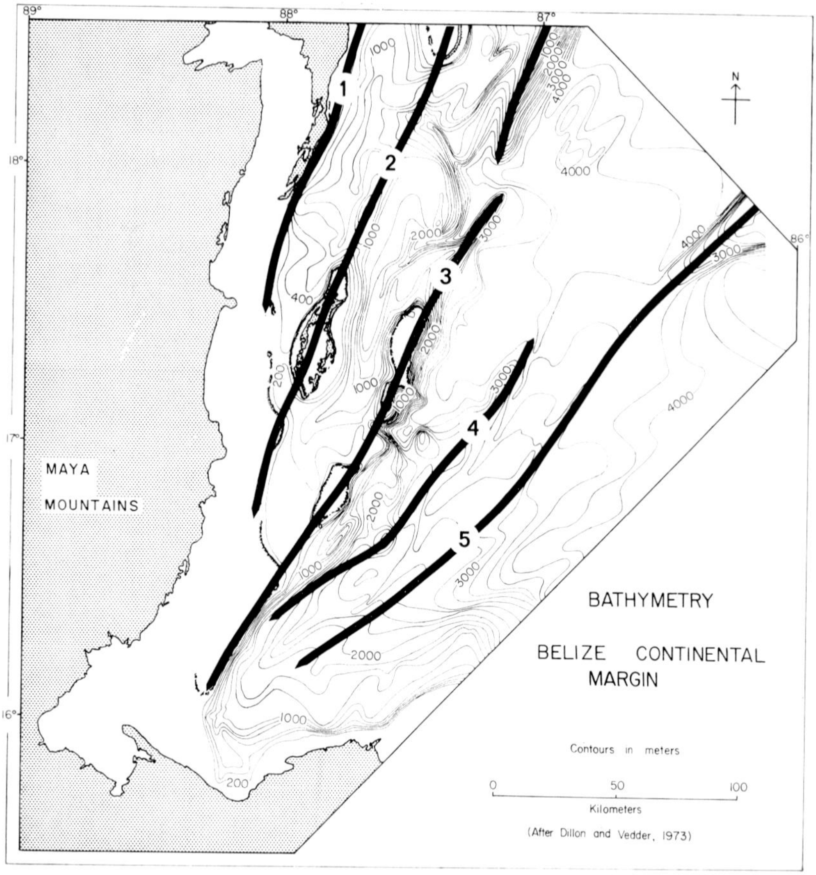

Fig. 1-8. A bathymetric chart of the western Caribbean Sea off Belize with the main submarine ridges highlighted (*after* Dillon & Vedder, 1973).

The best developed ridge forms the southern margin of the continental shelf and the basement on which Glovers Reef and Lighthouse Reef are located. The eastern margin of this ridge is a scarp with more than 3000 m of relief (Figs 1-8, 1-9). Seaward and in much deeper water are two other ridges, an 'outer basement ridge' (No. 4, Fig. 1-8) that is covered by sediments up to 200 m thick (Dillon & Vedder, 1973) and, to the east, the Cayman Ridge (No. 5, Fig. 1-8). Both ridges are characterized by local highs. Landward and almost parallel with the Glovers Reef–Lighthouse Reef structure is a ridge (No. 2, Fig. 1-8) that extends from Turneffe Islands to Banco Chinchorro of Mexico, and northward is fragmented by faults. The least defined ridge (No. 1, Fig. 1-8) lies along the northern edge of the barrier reef and Ambergris Cay.

The acoustic basement of these ridges is thought to consist of volcanic rocks, granitic and intrusive rocks and upper Palaeozoic sedimentary rocks, phyllites and schists that crop out in the Maya Mountains. This conclusion is borne out by recent drilling on Glovers Reef, Turneffe Islands and Ambergis Cay, all of which bottom in similar lithologies such as phyllites and extrusive intermediate to mafic volcanic rocks (Dillon & Vedder, 1973).

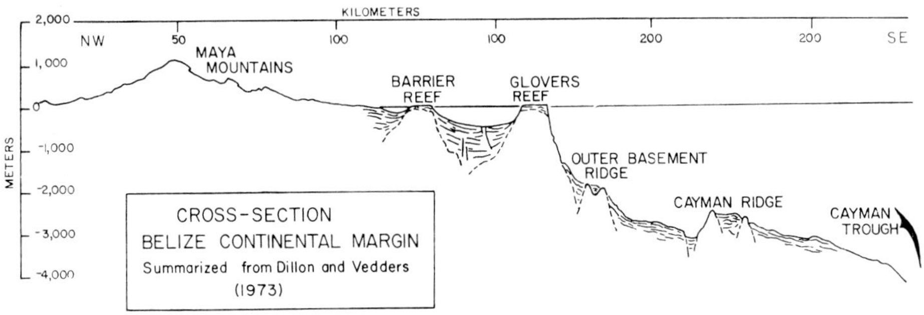

Fig. 1-9. A cross-section through the Belize continental margin, from the Maya Mountains seaward across Glovers Reef; compiled from data in Dillon & Vedder (1973) and topographic maps.

Dredging along the Yucatan Channel to the north of Isla Cozumel recovered samples of phyllite and marble (Vedder *et al.,* 1973). Even the Cayman Ridge is originally of continental origin as dredge hauls have recovered a sequence of rocks comprising granodiorite at the base and basalt at the top (Fox & Schneiber, 1970). On Glovers Reef 570 m of reef limestone were recorded above acoustic basement in drilling. Dillon & Vedder (1973) point out that much of the relief on other ridges may also be due to coral reef growth and be indistinguishable from underlying rock because of similar high seismic velocities and lack of coherent reflectors.

Recent detailed geophysical surveys (Dillon & Vedder, 1973) illustrate that these ridges that, on the basis of drill-hole data, have been emplaced since at least the Palaeocene have acted as dams, with sediment being deposited between ridges and occasionally lapping over the top (Fig. 1-9). These ridge-dams have prevented terrigenous clastic sediments from moving seaward. The basinal sediments, dominantly biogenic carbonates, have clearly been disturbed by slumping, probably from submarine erosion and faulting.

Structural evolution

The continental margin of the Belize and adjacent Yucatan appears to be the rifted trailing edge of a plate boundary formed by the opening of the Yucatan Basin (Uchupi, 1973; Dillon & Vedder, 1973). This model assumes that the Yucatan–Guatemala–Honduras block has always been attached to North America and was not, as Freeland & Dietz (1971) earlier suggested, located formerly in the Gulf of Mexico. While both Uchupi (1973) and Dillon & Vedder (1973) agree that the Yucatan basin and Cayman Trough were formed by two separate sets of plate movements, they disagree as to the timing of these events. Uchupi (1973) proposes: (1) a counter-clockwise rotation of Yucatan from late Mesozoic to early Cenozoic along faults south of Yucatan forming the Yucatan basin and (2) later rifting during the late Cenozoic leading to the formation of the Cayman Trough. Dillon & Vedder (1973), however, suggest that initial separation began earlier, with (1) sphenocasmic opening of the Yucatan basin north of the Cayman ridge pivoting around the southwest corner of the area in the Triassic to Jurassic, and (2) later movement of the Caribbean plate relatively eastward and away from the North American plate, beginning in the late Cretaceous, to form the Cayman Trough.

Dillon & Vedder (1973) further suggest that the onset of plate movement in the late Cretaceous induced uplift and thus a source for the shales, siltstones, and sandstones that were deposited in a trough that trended across the shelf out of the Gulf of Honduras and were encountered in wells both on the Turneffe Islands and Glovers Atoll. With continued opening of the Cayman Trough the fault blocks subsided and reef growth continued on them (e.g. 570 m on Glovers Reef, 1000 m on Turneffe Islands). The offshore Tertiary history is characterized by reef growth on fault block highs and deposition of sediment between ridges. Movement is continuing today as evidenced by modern seismic activity in the Cayman Trough (Sykes & Ewing, 1965; Molmar & Sykes, 1969; Bowin, 1968).

Other scattered evidence such as (1) faults that are seen to break the surface in geophysical profiles (Dillon & Vedder, 1973), (2) tilted beachrock on Turneffe Island (Stoddart, 1962), (3) tilted Pleistocene stalactites in submerged caves on Lighthouse Reef (Dill, 1971a, b), (4) earthquakes on the continental margin (Dillon & Vedder, 1973) and (5) faults in eastern Yucatan that cut rocks as young as late Tertiary (Viniegra, 1971), all suggests that the continental margin is still in the process of development.

SUMMARY

The reefs of Belize represent the largest reef complex in the Atlantic–Caribbean area and, with the presence of three different offshore atolls, rival the much larger Great Barrier Reef in the complexity of coral reefs and variety of sediment types. The complex lies in an open ocean setting, established on a block-faulted, trailing continental margin where the patterns of sedimentation have been more or less constant since early Tertiary time.

The Holocene record of sedimentation on the shallow shelf is now well known and reflects the gradual rise of sea level accompanied by a transition from siliciclastic to carbonate sedimentation. There has been a significant amount of Holocene accretion on the shelf, the style of which has been controlled to a great degree by pre-existing Pleistocene topography. The present sediments, particularly on the southern part of the shelf, have a distribution similar to those found on other modern tropical shelves that border continental masses (Ginsburg & James, 1976): terrigenous sediments adjacent to the mainland, transitional terrigenous to carbonate sediments in the centre of the shelf, carbonate sediments on the outer part of the shelf. A maze of isolated patch and platform atoll reefs occur in the seaward half of the southern shelf and in the lagoon of Glovers Atoll, while the margins of both are formed by an almost continuous barrier reef. Sediments along the margin are predominantly *Halimeda* sands with sediments in and around the reef crest rich in coral and coralline algae particles. Reef-crest derived sediments are prograding shelfward except around passes through the barrier where they are being funnelled basinward.

The fore-reef zones of this complex are either a gradual reef-to-basin transition similar to many ancient carbonate complexes or a steep reef-to-trench transition, with both types present off the southern part of the shelf and around Glovers Reef.

Chapter 2

The geophysical anatomy of the southern Belize continental margin and adjacent basins

PAUL ENOS*, W. JERRY KOCH† *and* NOEL P. JAMES

**Department of Geological Sciences, State University of New York at Binghamton, Binghamton, New York* 13901, *and*
*†Yankee Hill Exploration, R.R.*1, *Box* 604, *Idaho Springs, Colorado* 80452, *U.S.A.*

INTRODUCTION

To put the study of surface and near-surface submarine geology off the southern barrier reef and Glovers Reef in perspective, a detailed seismic survey was carried out in the same area. Seismic reflection profiling (Fig. 2-1) with a 9000 Joule sparker produced 230 km of records showing large-scale features. Two styles of profiling were carried out: (1) a close grid network of profiles adjacent to the barrier reef margin, normal and parallel to the trend of the structure (Figs 2-3, 2-4); (2) long lines across the trough between the barrier reef and Glovers Atoll and further seaward across the submarine ridges on the slope into the Cayman Trough (Figs 2-5, 2-6, 2-7, 2-8).

METHODS

Navigation depended on radar triangulation within range of land (about 32 km) and dead reckoning elsewhere.

The sparker generated a maximum of 9 kJ with two sound sources. Sub-bottom penetration was normally 0·25–0·50 s, two-way time. The peak frequency was about 240 c.p.s., giving a theoretical minimum bed resolution of about 1·5 m (1/4 wave length using salt-water sound velocity of about 1500 m s^{-1}). The characteristic reflection waveform contains 5·5 or 6 cycles extending over 25 m s^{-1}, or about 20 m of record at sound velocity in water, corrected for two-way travel time. This tended to obscure more closely spaced reflectors.

Within these limitations an attempt was made to recognize all coherent reflections. All wave trains that exceeded the characteristic 6 cycles were examined for divergence of lines within the wave train locally or when traced for long distances. This indicates that the longer wave train results from two or more reflections partially superposed

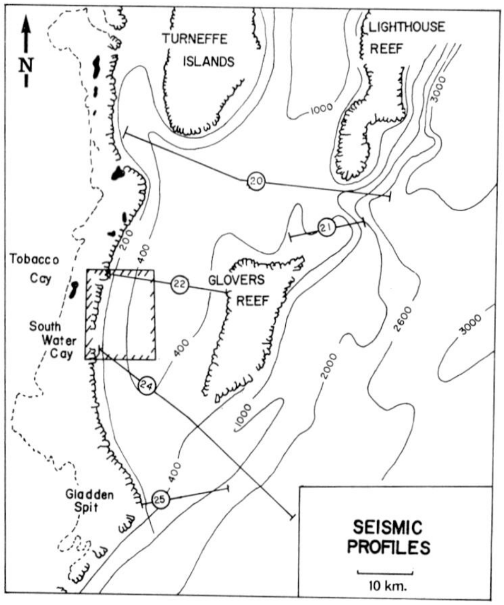

Fig. 2-1. A chart of the southern portion of the Belize barrier and atoll reef complex with the seismic lines illustrated in this chapter plotted and numbered; depth contours in metres. The box outlines the area of detailed profiling along the barrier reef and is enlarged in Fig. 2-2.

rather than from internal multiples (reverberations). Where feasible, the top of the lower, superposed, reflector was reconstructed by measuring the interval appropriate for 6 cycles (25 m s^{-1}) above the lowest line of the reflection signature. Otherwise a 'form line' was drawn on a persistent line within the signature. No corrections were made on the tracings for slope migration, velocity gradients, or slight variations in boat speed.

Locations on profiles cited in the test are given by profile number, kilometres along the profile (scaled at the top of each tracing), and travel time below sea level (left hand scale).

WALL-TO-BASIN TRANSITION

A grid pattern of closely spaced seismic lines comprising four east–west profiles, three oblique profiles and two north–south profiles was run over a distance of up to 6 km from the reef wall between Tobacco Cay and South Water Cay (Fig. 2-2). This is the same area examined in detail using the submersible and piston cores.

The subhorizontal, coherent reflectors that characterize the basin between Glovers Reef and the barrier reef do not extend to the foot of the wall but grade into a seismic 'transparent zone' some 2–3 km in front of the wall (Figs 2-3, 2-4). The contrast between the 'transparent zone' and basinal reflectors is demonstrated by comparing profile 1 along the base of the escarpment with few coherent reflectors and a parallel profile 2 run some 4 km further seaward (Fig. 2-3). The 'transparent zone' does not extend to the present sea floor but is overlain by reflecting horizons about 25 m thick at the basinward end, 50 m thick in the middle and 75–100 m thick at the top of the

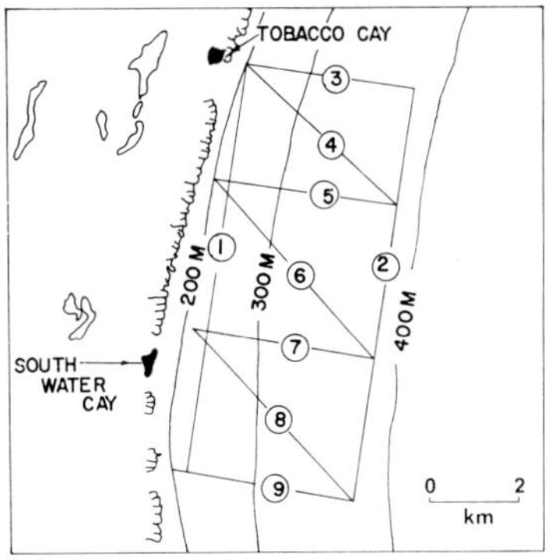

Fig. 2-2. The area between Tobacco Cay and South Water Cay along the barrier reef in which a grid of seismic transects were run; depth contours in metres. The lines in this grid are illustrated in Figs 2-3, 2-4 and 2-5.

sediment wedge at the foot of the wall (Fig. 2-4). The top of the 'transparent zone' is closer to horizontal than the overlying sediment surface but has an irregular upper surface that commonly displays terrace-like relief, stepping down basinward (Fig. 2-4, legs 5, 6 and 8). The base of the 'transparent zone' is indicated by vague deeper reflections in some profiles (Fig. 2-4). The lateral transition to good basinal reflectors and underlying reflecting horizons indicates that the zone is not simply a high in acoustic basement.

Equivalent 'transparent zones' are apparent at the base of the barrier reef in profiles 22 and 24 to the north and south (Figs 2-5, 2-6), but are lacking in profile 25 (Fig. 2-8) further south along the barrier reef. 'Transparent zones' in equivalent positions but much reduced in size are seen at the base of the leeward escarpment of Glovers Reef in profiles 21 (Fig. 2-8) and 24 (Fig. 2-6).

We interpret these 'transparent zones' either as zones of debris from oversteepened reef escarpments or as reefs and reef-associated deposits formed during Pleistocene low stands of sea level.

If they are zones of debris, then the lack of internal layering and coarse particle size with high porosity would account for the seismic transparency. It is possible that this debris accumulated particle by particle as a fore-reef talus, but the lack of a steep slope on top of the wedge and lack of internal layering suggest mass movement. Mass movements may have been entirely submarine as excessive reef debris accumulated in unstable positions in shallow water, or may have occurred when low stands of sea level exposed at least the upper part of the escarpment.

During Pleistocene interglacial periods world-wide sea level dropped to at least 120 m (Milliman & Emery, 1968) and possibly more for extended periods of time. During these periods reefs certainly flourished along the sides of the basin at the foot of steep limestone cliffs. If conditions were similar to those of today, then reef growth extended at least 70–80 m below the sea level stand of 120 m and debris shed from these reefs 100 m or more below that. Alternatively we suggest that these transparent zones might be reefs and reef-associated deposits developed during Pleistocene glacial intervals and now buried by late Wisconsin and Holocene marginal sedimentation.

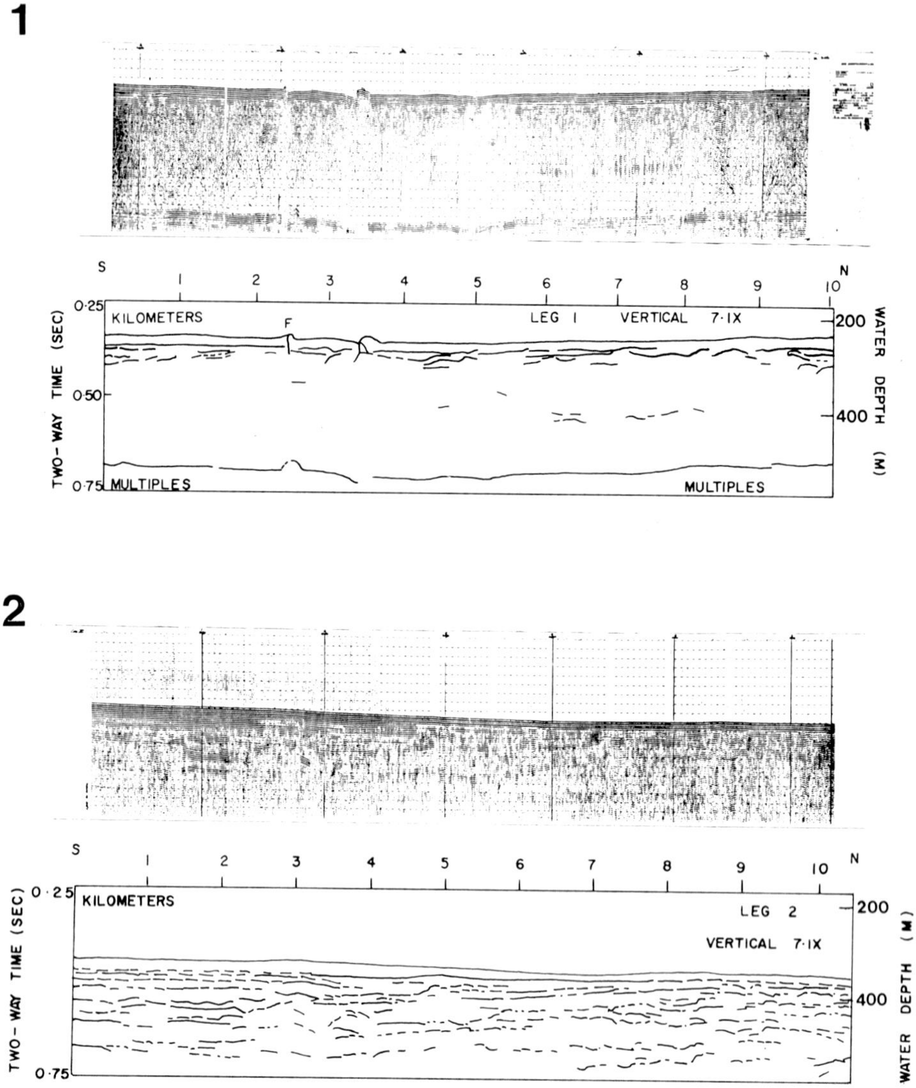

Fig. 2-3. Seismic lines 1 and 2 (see Fig. 2-2) with the actual record and a tracing of the major reflectors (slightly enlarged) illustrated for each line; F, fault.

CONTINENTAL SLOPE AND SHALLOW BASIN SEDIMENT PACKAGES

Geometry of sediment packages

Offshore ridges appear to have influenced sedimentation by forming basins on their landward or shallower sides as well as providing platforms for reef growth on their crests (Dillon & Vedder, 1973). Sediment packages in basins behind ridges are generally lens-shaped, thickening in the centre of the depression. Present day surfaces of sedimentation are either concave (Fig. 2-5) or almost flat (Fig. 2-6).

Concave profiles are bounded by prominent reef-capped ridges at either end. The

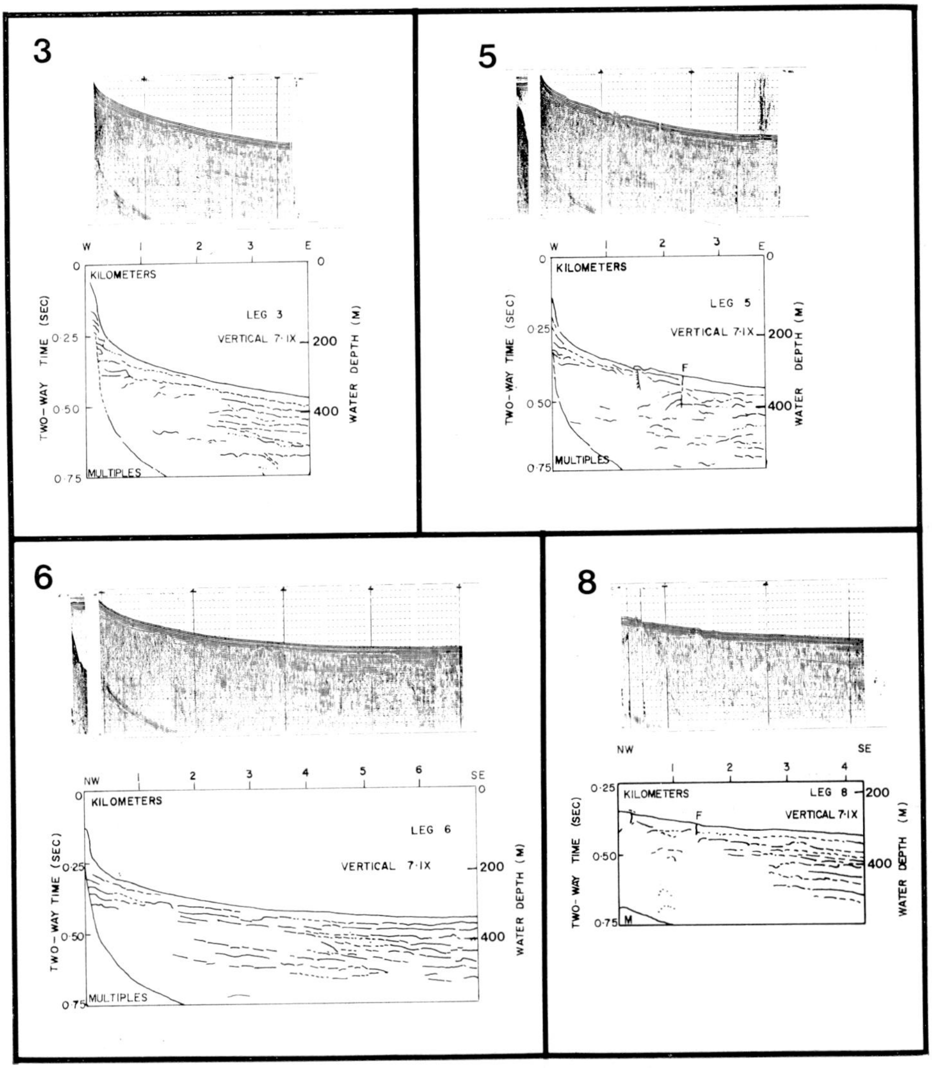

Fig. 2-4. Selected seismic lines 3, 5, 6 and 8 (see Fig. 2-2) with the actual record and a tracing of the major reflectors illustrated for each line; F, fault. The break near the left margin in actual records 3 and 5 is a change in the recording interval from 0 to 0·5 s at the left (note the direct wave) to 0·25–0·75 s.

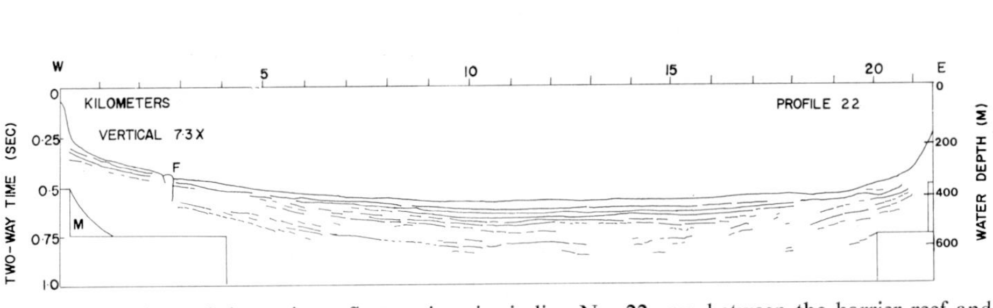

Fig. 2-5. Tracings of the major reflectors in seismic line No. 22, run between the barrier reef and Glovers Reef (see Fig. 2-1).

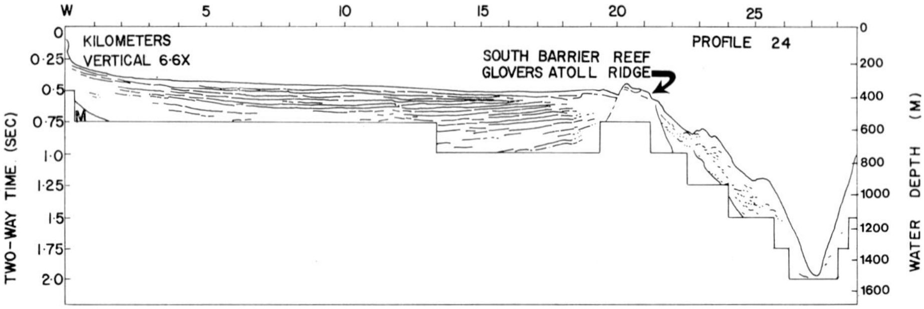

Fig. 2-6. A tracing of the major reflectors in seismic line No. 24 (see Fig. 2-1), run from the barrier reef southeast across the offshore ridge (number 3, Fig. 1-8) on which Glovers Reef is located.

shape probably results, not from 'damming', but from uniform pelagic sedimentation in the basin centre, as indicated by cores, augmented by significant input of shallow-water detritus at the basin margins as indicated by samples and seismic records (see Chapter 4).

Flat profiles are established where no ridge existed, where the ridge is covered by sediments, or where the ridge is deeply submerged. In these profiles, however, the sub-bottom reflectors are concave up, producing the lens-shaped package. Either (1) flat profiles developed from concave profiles as the ridge was covered by sediment or as it became too deeply submerged to produce sediment and supply debris, or (2) if flat profiles remained flat during sediment accumulation, lens-shaped packages reflect more rapid subsidence of the basin centre. If this latter postulate is the case, the flat profiles must represent submarine profiles of equilibrium to compensate for the differential subsidence.

Turbidity currents or other bottom currents might produce such a profile parallel to the flow axis but would probably result in channels and levees visible in bathymetric profiles intersecting the axis of flow. No such features were seen at the present sediment surface in 230 km of profiles.

Correlation of reflecting horizons within the sediment packages between profiles does not appear feasible. Even within the close grid of profiles 1–9 reflectors cannot be correlated very satisfactorily, partly because reflectors were consistently lost in the transparent zone at the base of the barrier reef escarpment. In adjacent profile 22 (Fig. 2-5) the character of sub-bottom reflectors changes with more wavy, less persistent layering at depths greater than 120 ms below the sediment surface. Equivalent changes are not seen in other profiles (contrast Profile 25, Fig. 2-8).

Slumps

Masses of material forming topographic highs on the sea floor, illustrating chaotic internal reflections, and in intimate association with escarpments of considerable relief suggest episodes of submarine slumping. These slumps are well illustrated in the northern extension of the trough that separates the barrier reef and Glovers Reef (Fig. 2-7).

Small escarpments at 20 and 22 km east of the Lighthouse Reef Ridge are probably slump scars, with the slumped masses just to the east of each. Good reflectors beneath

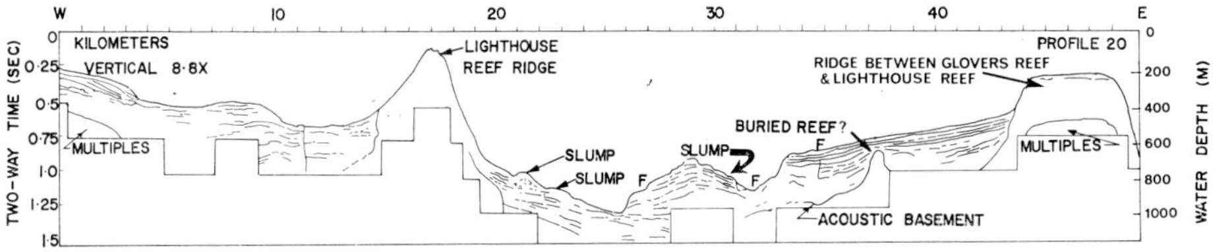

Fig. 2-7. A tracing of the major reflectors in seismic line No. 20 (see Fig. 2-1) run from the barrier reef through the gap between Glovers Reef and Lighthouse Reef and so crossing both the ridges on which Turneffe Islands (No. 2, Fig. 1-8) and Glovers Reef–Lighthouse Reef (No. 3, Fig. 1-8) are located.

these lenses suggest that the soles of the slumps are 15 ms (at 20 km) and 9 ms (at 22 km) beneath the sediment surface.

The 200 m high escarpment at 33 km has no obvious cause such as faulting or reef growth. Acoustic basement (heavy line at 1·2–1·3 s) is displaced little, if any, at the escarpment. The huge package of sediment to the west (26–32 km) which produces 280 m of sea-floor relief is probably related to the escarpment. The internal structure of the package is well layered and coherent, but reflectors are convex up and distorted by numerous small faults, which break the surface. This package overlies nearly horizontal reflectors at 1·28 s.

This package was probably emplaced by mass movement of approximately 1 km away from the escarpment (km 33) in which the trailing edge rotated down and the crest bulged upward by internal flowage. Internal cohesion was maintained except as reflected by the faults. Westward dip of the western edge of the package may reflect original dip toward the deeper floor of the basin. The origin of the smaller escarpment at km 26 is unsolved; it may have been the edge of a prograding package like that in profile 25 (km 0–4) which became oversteepened to produce the slump. The escarpment may also be related to the descent of acoustic basement from 1·2 s at 32 km to some depth greater than 1·5 s at km 28 (Fig. 2-8). If this drop in acoustic basement is caused by a major fault somewhere between km 28 and 31, it may have formed a west-facing escarpment at the surface which later slumped, and moved westward over 2 km.

Faults

In addition to the major faults that bound ridge structures, stratification in the sediment packages is often broken by smaller faults. These small faults, offsetting the present sediment surface, are most common towards the sides of the basins and can be seen in profiles 1 (Fig. 2-3), 5, 7 (Fig. 2-4), 20 (Fig. 2-7) and 22 (Fig. 2-5). These faults appear to be local, although one small fault may be traceable through profiles 1, 7, and 8 (km 2·4, 0·1, 0·2 respectively). Small faults that do appear on adjacent profiles often have a different sense of movement. Side reflections with up to 20 m relief in profiles 5 (1·5 km), 8 (0·2 km) and 1 (3·5 km) may indicate large blocks on the sea floor with peripheral sediment scour, rather than faults, especially since they are near the barrier reef escarpment.

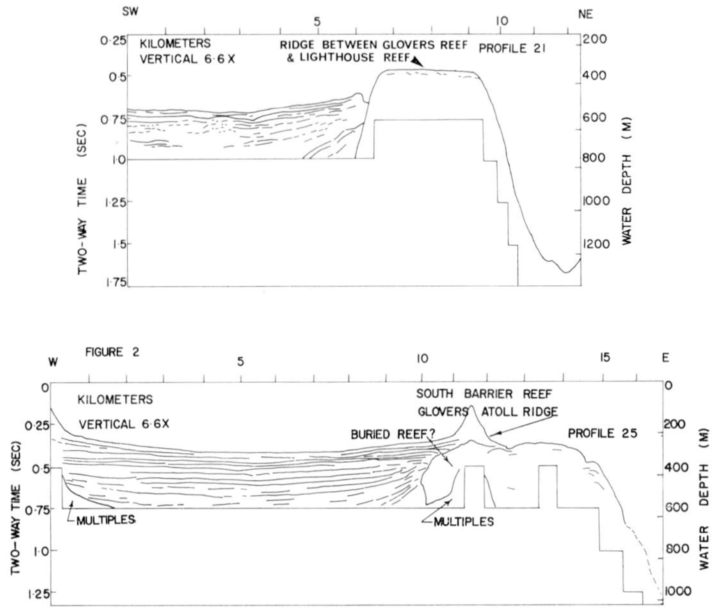

Fig. 2-8. Tracings of the major reflectors in two seismic lines run across the southern barrier reef–Glovers Reef ridge (No. 3, Fig. 1-8); see Fig. 2-1 for location.

ORIGIN OF SUBMARINE RIDGES

Dramatic breaks in slope along the Belize margin are formed by the submarine ridges on the trend of Turneffe Islands and the barrier reef north of Tobacco Cay and on the trend of Lighthouse Reef, Glovers Reef, and the barrier reef south of Gladden Spit (Fig. 1-8). These obstructions, which form the lips of small basins, have been referred to as 'narrow, sediment-damming ridges', and 'tectonic dams' by Dillon & Vedder (1973). Interpretation of these ridges as tectonic in origin, presumably as horst blocks, is supported by their linear trend and by crystalline rocks dredged from the seaward-facing escarpments further north (Chapter 1). It is consistent with the steep, straight sides of the ridges seen in sub-bottom profiles (e.g. Profile 25, Fig. 2-8) and with the general lack of internal reflections, suggesting deformed or non-stratified rocks.

A further indication of structural origin is the internal truncation within the sedimentary pile adjacent to the ridges (Figs 2-7, 2-8), which suggests reactivation of the faults during sedimentation, probably during the Tertiary.

Several lines of evidence suggest that the ridges have been considerably modified by reef growth or other sedimentation.

The three 'atolls' or shallow water reef complexes that sit atop the ridges (Turneffe, Glovers and Lighthouse) are 11–15 km wide at sea level, contrasting with a maximum width of 5 km for the submerged ridge crests even in oblique crossings. This widening of the ridges may be ascribed to irregular fault planes, but, in view of the general linearity of the ridges, it would appear that accretion by reef growth and shallow-water sedimentation is a more reasonable explanation.

A more extreme example of modification of structural trends by sedimentary or biological processes is the construction of the barrier reef. Projections of the barrier reef at Gladden Spit and north of Tobacco Cay align with the structural grain. In the arcuate re-entrants between Gladden Spit and Tobacco Cay and west of Turneffe Islands, reef growth has maintained sedimentation near sea level despite being on a structurally low area. This may have been possible because the regional structural dip brings interhorst areas generally closer to sea level at this latitude or because overall structural relief is less.

Ridge flanks flatten with depth in several sub-bottom profiles (Figs 2-6, 2-7). Normal faulting is unlikely to produce such flattening. Sedimentary processes such as expanded reef growth or accumulation of sediment wedges provide a more satisfactory explanation.

The general lack of internal reflectors within the ridges, while consistent with a structural origin, would also be expected from reef construction with a rigid framework, lacking internal layering and enclosing high internal porosity. A veneer of sediment on top of the ridges is suggested in Figs 2-6, 2-7, 2-8.

Finally, the effectiveness of reef growth in producing relief is suggested by two very probable reef-growth features in profiles 20 (Fig. 2-7) and 25 (Fig. 2-8). The buttress in profile 25 (Fig. 2-8) appears to be constructional relief, merging with sediments to the west and overlying a surface of contrasting lithology at 0·34 s.

The buried feature in profile 20 (Fig. 2-7) is probably not a diapir because overlying sediments are undisturbed. This fact also argues against a horst origin, as does the merger of the base of this feature with the nearly horizontal acoustic basement to the west. The dip of the sediments adjacent to the feature may be initial dip of reef debris or post-depositional draping over the rigid buttress. The location of this postulated reef is a bit puzzling. Nothing is known of its three-dimensional shape. It may have been a patch reef behind the ridge or a linear reef rimming the basin before development of the ridge at the present shelf edge by faulting and/or reef growth. The layer which reflects uplift on the ridge nearly covered the postulated reef prior to uplift.

SUMMARY

Seismic 'transparent zones', which may be wedges of debris from oversteepened margins above or buried Pleistocene reef and reef associated deposits formed during low stands of sea level, are common beneath 50–100 m of sediment adjacent to the wall. Whatever their origin these structures contribute significantly to the wedge-shape geometry of the fore-reef, especially along the barrier reef.

Large linear submarine ridges, the major structural elements in the area, were formed by normal faulting, as suggested by Dillon & Vedder (1973), but further seismic profiling indicates that they have been considerably modified by carbonate accretion, particularly reef growth. Possible buried reefs can also be recognized adjacent to one of the major ridges and in deep water, buried by later basinal sediments.

Basins between ridges are partially filled by lens-shaped packages of mainly pelagic sediment. Sediment packages in deeper water basins with rugged topography illustrate large slump or mass movement structures. The upper several hundred metres of pelagic sediments are locally cut by faults, particularly towards the basin margins.

Chapter 3

The morphology, sediments and organisms of the deep barrier reef and fore-reef

INTRODUCTION

The southern part of the Belize region (Fig. 3-1) exhibits two styles of platform margin: (1) reef to shallow (400 m) basin, and (2) reef to deep oceanic trough. Using these two settings and the variations in reef type between barrier and atoll, we selected seven sites for study. For the reef to shallow basin variety, we examined three localities on the barrier reef and two on the leeward or western side of Glovers, Sites 1, 2, 3, 5 and 6 (Fig. 3-1). As examples of the reef to oceanic trough, we chose two positions, one along the southernmost part of the barrier reef, Site 4, and the other on the seaward side of Glovers Reef, Site 7 (Fig. 3-1).

At each of these seven sites we described and sampled the fore-reef to depths of 300 m from a research submersible. This chapter presents summaries of our observations on the morphology, the sediment types, and the organisms. We did not spend an equal amount of diving time at each of the seven sites, but instead, as indicated in Table 3-1, we concentrated on two sites believed to be representative of the two

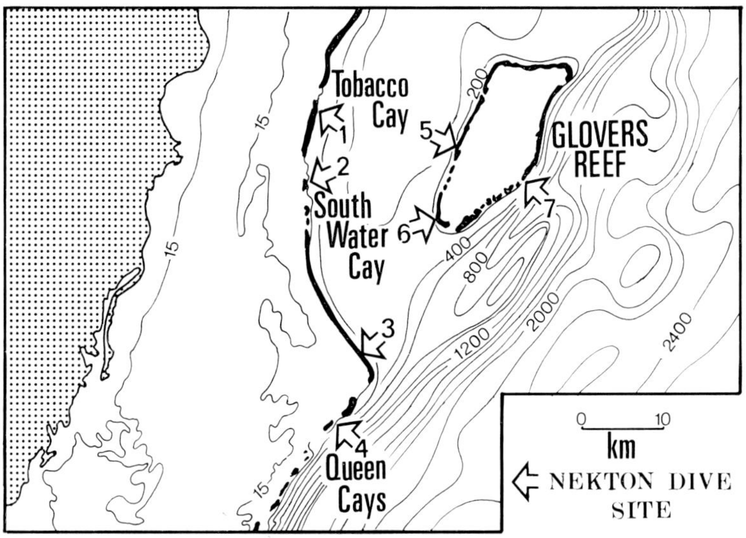

Fig. 3-1. The location of NEKTON dive sites (numbered with arrows) in the area of detailed study in the southern part of the Belize barrier and atoll reef complex.

major settings. Using the accumulated experience from these two sites, we were able to evaluate the similarities and the differences between each of them and the other five sites.

FIELD METHODS (OPERATIONS)

We used the free-diving submersible NEKTON* to examine and sample the fore-reef zone to depths of 300 m. The NEKTON is 4·5 m long; it carries a pilot and observer who can see and photograph on either side through portholes. NEKTON is equipped with a depth gauge, an inclinometer, lights, and a strobe mounted externally for viewing and photography. It has a tong-like manual claw that the observer uses to pick up specimens that are stored in a retractable canvas bag. Because of the heat and humidity, individual dives were usually less than 90 min, and for reasons of safety, diving was limited to daylight hours. With these restrictions and the time required to launch and retrieve, we were able to dive from 8 to 10 h per day.

To explain how the submersible was used to describe and sample the fore-reef zone it is convenient to give an account of the various steps beginning with the locating of the transect and ending with the collection of samples.

Specific sites were located using bearings on islands, or by dead reckoning from landmarks. Once a line of transect was chosen, an anchored buoy was set near the major break in slope as a reference point. An initial dive was made to reconnoitre the fore-reef, and it was usually followed by a series of dives in which the observers and pilots concentrated on describing the major topographic features, noting the principal animals, measuring the slopes, and photographing representative features. From these descriptions, sites were chosen for the collection of sediment and rock samples.

Observations by the two-man crew were recorded on a cassette tape recorder and they were keyed to depths indicated on the gauge mounted in the submersible. Photographs were taken in colour and black and white through the portholes with 35 mm reflex cameras using an external strobe or the submersible running lights for illumination at distances less than 3 m, and ambient light for greater distances. A summary of the diving time is given in Table 3-1.

TERMINOLOGY

Because this monograph treats not only present-day morphology, but organisms and sediments as well as the rock record of these modern elements, we have tried to choose terms that would be compatible with the existing terminology of both fossil and recent reefs and also carbonate platform margins.

In the description of ancient reef complexes it has become standard practice to separate the marginal zone into two facies belts, the *reef facies* and the *fore-reef facies* (Henson, 1950; Link, 1950; Cloud, 1952; Zankl, 1969; Playford, 1969; Krebs & Mountjoy, 1972) and this separation extends to reefs of Pleistocene age (Mesolella, Sealy & Matthews, 1970; James, Stearn & Harrison, 1977). The *reef* or *reef-core facies* is the in-place accumulation of large skeletal metazoa and derived sediment characterized in the rock record by biolithites (Folk, 1962) or framestones to bindstones

* Owned and operated by General Oceanographics, Inc., 11578 Sorrento Valley Road, San Diego, California 92121.

Table 3-1. Operations summary*

Site no.	Site	Number of observers	Number of dives	Total hours underwater	Hours of description and photography	Hours setting explosives and collecting samples
Barrier reef						
1	Tobacco Cay	7	38	50·0	21·0	29·0
2	South Water Cay	3	4	4·5	2·0	2·5
3	Buttonwood Cay	1	1	1·1	1·1	0·0
4	Queen Cays	3	6	12·3	10·2	2·1
Glovers reef						
5	Western Lee	3	4	3·4	3·4	0·0
6	Southwestern Lee	4	4	5·3	3·5	1·8
7	Eastern	9	33	24·8	18·5	6·3
Totals			90	101·40	59·7	41·7

* 1971 and 1972.

(Embry & Klovan, 1971), often with demonstrable relief above surrounding bedded sediment. The *fore-reef facies* generally separates reef and basin deposits and is composed of sediments that are clearly derived from the reef and deposited on a surface that slopes basinward.

Since not all carbonate banks or 'buildups' are reef-fringed, the term fore-reef is not applicable to many situations in the rock record, and so the term *foreslope* (Wilson, 1969) has evolved as a more general term for the same facies. Davies (1977) has succinctly described the *shelf foreslope* as the narrow zone seaward of the shelf edge characterized by steep depositional slopes and constructed largely of material derived from the shelf and shelf edge and the *slope* as a more extensive development of the shelf foreslope in which water depths increase more gently away from the shelf and rocks include well bedded dark-coloured carbonates and fine-grained turbidites.

There is general agreement on the terminology of major features to depths of 10 m or so on modern reefs; spur and groove, reef crest, etc. (see Battistini *et al.*, 1975). For the deeper zones, however, particularly below 30 m (the common depth limit of most SCUBA observations), there is considerable variation in nomenclature (Newell & Rigby, 1957; Logan, 1969; Rigby & Roberts, 1976; Zankl & Schroeder, 1972; Meischner & Meischner, 1977; Bak, 1977). There are well developed terms for this deeper zone from the much-studied Discovery Bay area in Jamaica (Goreau & Land, 1974; Lang, 1974; Land & Moore, 1977) but the use of the term fore-reef is at odds with its accepted meaning for ancient reefs and the term island slope is inappropriate for the margins of shelves or platforms.

In the broad view we have chosen to utilize the basic terminology established for the study of ancient reefs because we visualize the barrier reef as a constructional feature characteristic of rimmed shelves world-wide (Ginsburg & James, 1974). In Belize this often high-relief feature clearly separates the shelf lagoon from offshore basins and troughs. More specifically the barrier reef is that setting characterized by prolific coral and algae growth at the edge of the shelf or platform and as such would extend from the edge of the lagoon across the shallow water areas and down the seaward margin of the shelf to the lower limit of coral and calcareous green algae growth (here about 80 m) (Fig. 3-2). The fore-reef is that sloping part of the sea floor seaward of the zone of living coral and green algae growth characterized by reef-derived talus and sediment. Juxtaposed between the barrier reef and the fore-reef is a vertical escarpment, common to most modern platform margins but rare in ancient examples, which we have termed the wall. Because corals and calcareous green algae grow on the upper part of the wall and it is obviously different from the talus slope of sediment that comprises the fore-reef below, we have included the wall as the deepest and most basinward element of the barrier reef.

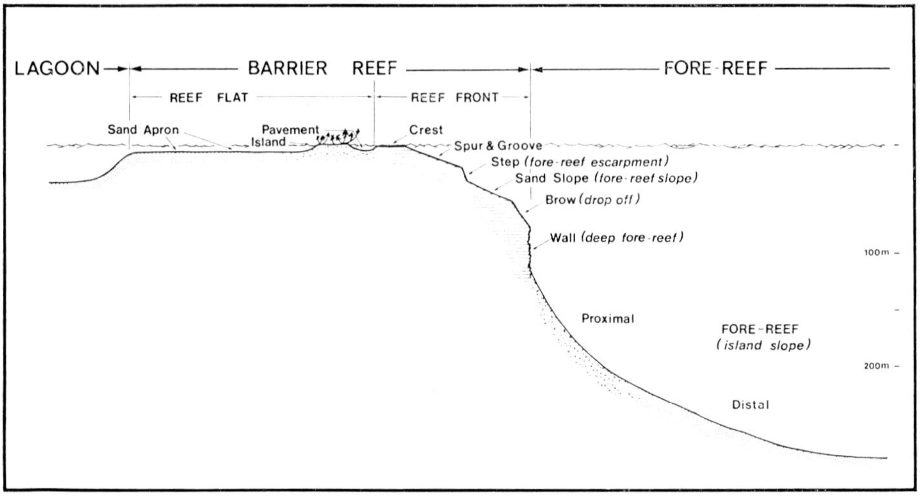

Fig. 3-2. A diagrammatic cross-section illustrating the different morphological elements on the seaward margin of the reef-rimmed Belize Shelf and atolls and the names used in this report. The terms in brackets are the names used by workers in Jamaica (Land & Moore, 1977) for similar elements.

The reef front as used here consists of smaller, but distinct subdivisions whose names are shown on Fig. 3-2, together with synonyms (in italics) used to describe similar features in Jamaica.

The nature of the sea floor seaward and basinward of the fore-reef is highly variable and has recently been documented by Schlager, Hooke & James (1976), Schlager & Chermak (1979) and Mullins & Neumann (1979). There appear, at present, to be three different styles of deep platform margins (Fig. 3-3), although these may be subdivided according to setting and local conditions: (1) the fore-reef may grade directly into a shallow trough, (2) the fore-reef is the upper part of a much larger, relatively gently dipping slope (20–40°) that grades into a deep trough or basin (called the *gullied slope* (Schlager *et al.,* 1976) or *slope* (Mullins & Neumann, 1979) in the Bahamas), and (3) the fore-reef forms the upper part of a very steep slope that borders a deep trough or basin, with little or no intervening slope.

In Belize we have studied the fore-reef in two of these situations. Between the barrier reef and Glovers Reef the fore-reef flanks a shallow trough and we have termed this style a sloping fore-reef. Along the southern part of the barrier reef and the eastern edge of Glovers Reef the fore-reef is the upper part of a series of cliffs and steep slopes that forms the western wall of the Cayman Trough and we have called this style a cliffed fore-reef.

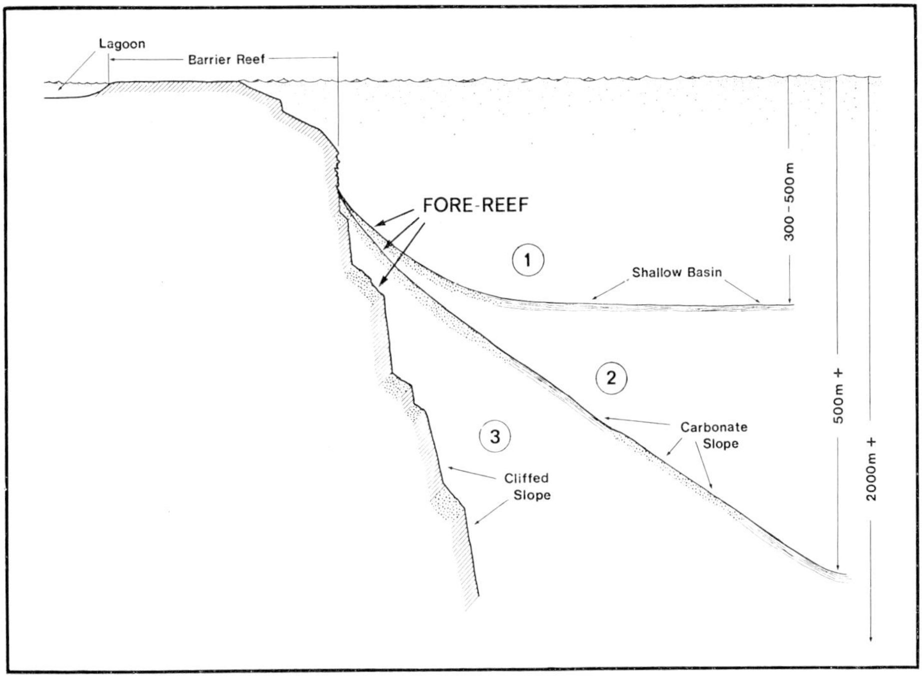

Fig. 3-3. A diagram illustrating three of the more common styles of reef-rimmed shelf-to-basin transitions described to date in modern oceans. See text for meaning of numbers. Not to scale.

Although it is easy to fix the upper limit of the fore-reef, it is more difficult to fix the lower limit. We have arbitrarily restricted it to that zone immediately adjacent to the wall which is characterized by sediments derived directly from the reef and the wall. The carbonate slope below the fore-reef is formed by resedimented deposits (turbidites, debris flows, etc.) derived from the barrier reef, the wall or the fore-reef alternating with pelagic carbonate muds.

VARIATIONS IN MARGIN MORPHOLOGY

Margins adjacent to a shallow basin

The five sites that border the shallow (400 m deep) trough, three along the barrier reef and two on the leeward side of Glovers Reef, have remarkably similar profiles (Figs 3-4, 3-5). There are, however, two kinds of second order differences that are seen by inspection: (1) the elevated ridge forming the brow in the profiles at South Water Cay (Site 2), and (2) variations in the continuity of the proximal fore-reef and in the depths of the boundaries of some elements. These variations are discussed in the sections summarizing the elements below.

The shallowest reaches of all profiles are formed by *spur and groove* structures that extend to an average depth of between 15 and 21 m. In most localities a vertical to near-vertical *step* extends below the spur and groove to a depth of 30–37 m. Only off South Water Cay is the step missing and the spur and groove extends to a depth of 33 m.

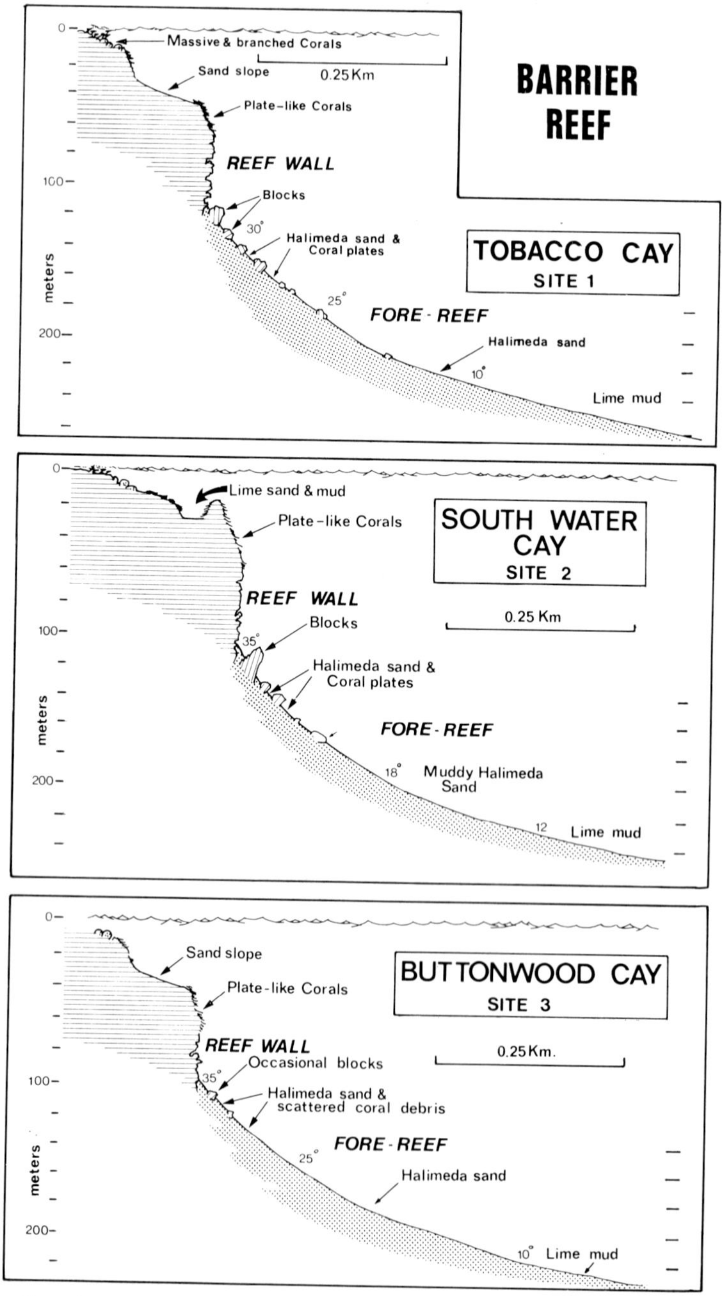

Fig. 3-4. Profiles of the shallow reef to basin transition at three different sites along the barrier reef (see Fig. 3-1 for location). These profiles are constructed from data obtained by depth soundings and direct observations from submersible.

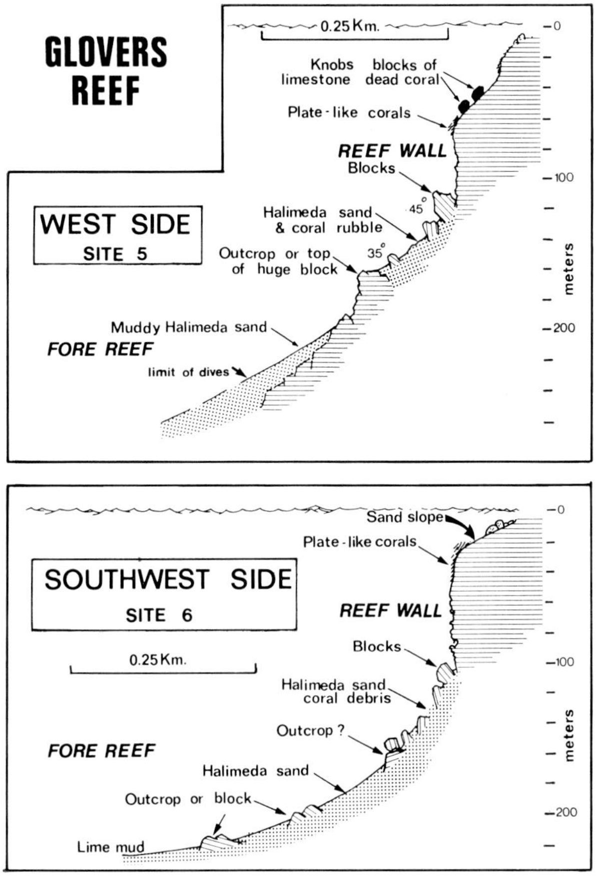

Fig. 3-5. Profiles of the shallow reef to basin transition at two different sites on the leeward side of Glovers Reef (see Fig. 3-1 for locations). These profiles are constructed from data obtained by depth soundings and direct observations from submersible.

The step is fronted by a *sand slope* that extends to a depth of 42–50 m where the bottom steepens abruptly (the brow) and passes into the top of the wall.

The wall extends from a consistent depth of about 65 m to the top of the sloping fore-reef at a depth varying from 105 to 116 m. The *proximal sloping fore-reef* is a jumble of talus blocks, coral debris and *Halimeda* sand with steep slopes (30–45°) near the wall which decrease rapidly seaward (Figs 3-4, 3-5), The lower limit of the proximal portion is taken at the places where slopes flatten to about 10°, because this position marks the change in sediment composition from sand to muddy sand and mud.* The *distal sloping fore-reef* is a monotonous, nearly flat sediment floor dotted with sediment cones of burrowing organisms that grades imperceptibly into the flat floor of the trough.

Margins along the Cayman Trough

The two sites along the northwest margin of the Cayman Trough, one at the

* Here and throughout the rest of this monograph, sand and mud refer to carbonate sand and carbonate mud, unless otherwise stated.

southern end of the barrier reef (Site 4) and the other on the eastern side of Glovers Reef (Site 7) are similar (Fig. 3-6). The shallow part of the profile at Queen Cays (Site 4) resembles others along the barrier reef described above but that on the eastern side of the atoll (Site 7) is different; the step and sand slope are missing, the spur and groove topography extends to a depth of 27 m and instead of a convex brow it is concave from 25 to 60 m (Fig. 3-6).

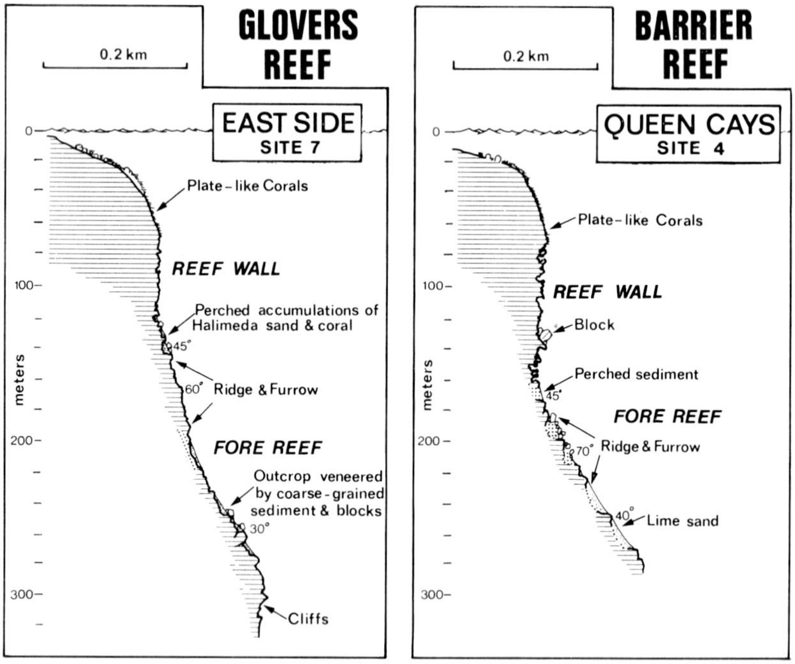

Fig. 3-6. Profiles of the upper 300 m of the margin of the Cayman Trough at two different sites, one on the barrier reef (Queen Cays) and the other on the eastern side of Glovers Reef (see Fig. 3-1 for locations). These profiles are constructed from data obtained by depth soudings and direct observations from submersible.

The deeper part of the profile, the cliffed fore-reef, is similar at both localities. Between 120 and 150 m the bottom is either an extension of the wall or steep cliffs alternating with talus accumulations of blocks, coral plates and sediments dipping from 30 to 70° seaward. Between 150 and 200 m the vertical to near-vertical cliffs are separated by gullied, sediment-covered slopes inclined at 50–70°. Below 200 m, there are fewer cliffs, they are not as steep and the slopes between them are wide, gently furrowed ramps dipping seaward from 30 to 45°. At site 7, the only place where we reached 305 m, the slopes end and the bottom drops away abruptly in another precipice.

MORPHOLOGY, ORGANISMS AND SEDIMENTS

The following section, which summarizes our observations on each of the physiographic elements, draws heavily on the Tobacco Cay (Site 1) and eastern side of

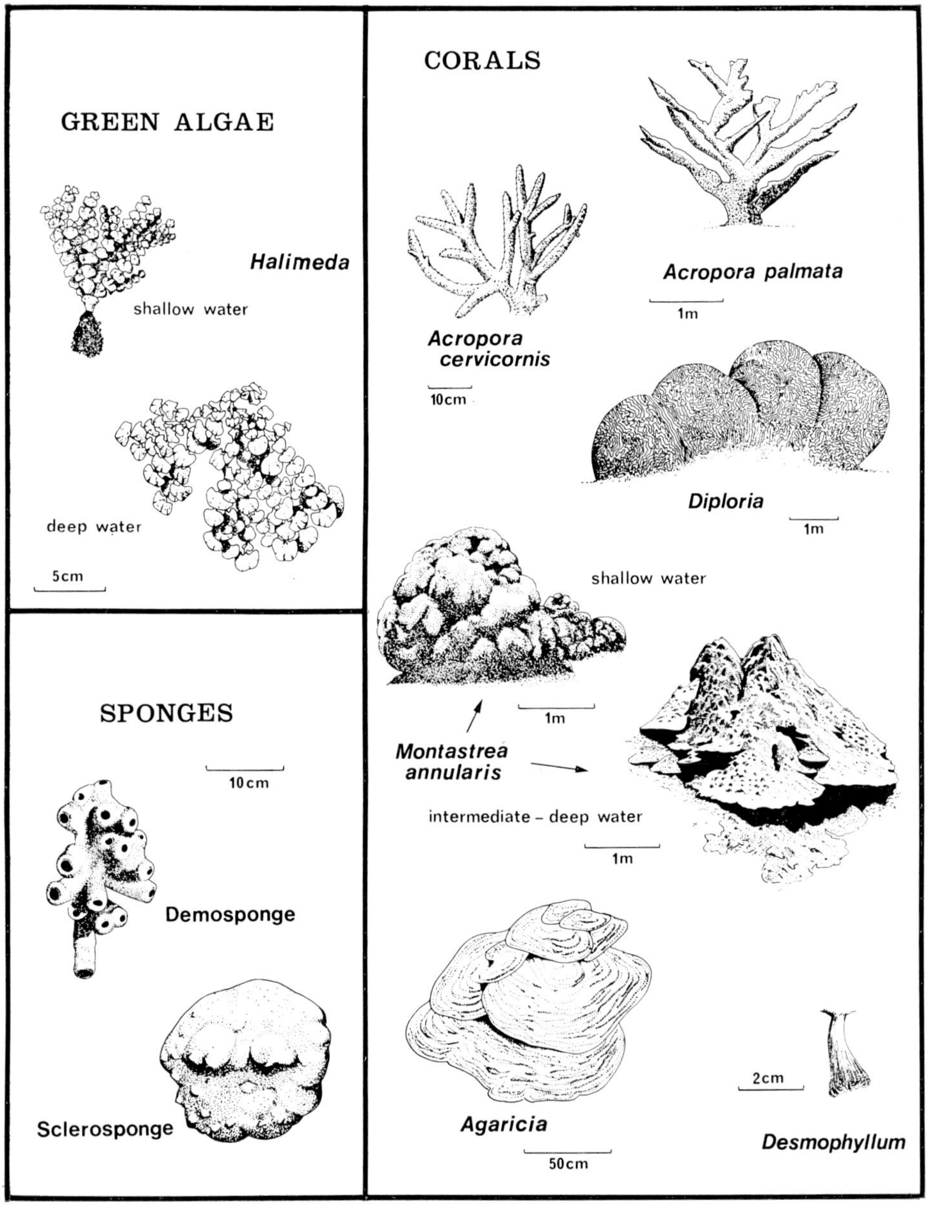

Fig. 3-7. A sketch of the most important biota, from a geological viewpoint, observed in submersible dives.

Glover's Reef (Site 7) where diving was concentrated (Table 3-1) but it includes observations from all sites. As a guide for readers unfamiliar with the Holocene biota, Fig. 3-7 illustrates the principal reef-building organisms: corals, green algae and sponges.

The spur and groove

The shallow spur and groove structure described in detail by James *et al.* (1976), so prominent at the reef crest and in very shallow water, can be traced seaward into deeper water as linear lobes of coral separated by sediment-floored grooves. Individual spurs are 4·5–6 m wide and are separated from one another by grooves that are up to 2·5 m wide and floored with both pebble to cobble size coral rubble and mottled, rippled, occasionally cross-rippled, sand. The crest of each spur is crowned with 15–40% living coral, mostly *Acropora cervicornis* and head-like corals, the most prominent of which is *Montastraea annularis.* Growing amongst corals are a profusion of sponges, alcyonarians, gorgonians, the most common of which are *Pseudopterogorgia* sp. and *Gorgonia ventalina*, and species of the green alga *Halimeda*, most of which are attached to a rock surface encrusted by coralline algae. The side of each spur at shallow depths is armoured with many overlapping plates of *Montastraea annularis*.

Below a depth of 15 m the amount of living coral begins to increase noticeably and the number of *M. annularis* colonies, in particular, begins to increase. Near the break in slope of the step, we estimate that three quarters of the surface of many lobes is covered with subcolumnar and plate-like colonies of *M. annularis.* At the top of the step, at 20–21 m, these overlapping plates of *M. annularis* as well as *Agaricia* form irregular to circular mounds resembling beehives, 4–5 m high. Relief between spurs and grooves at the top of the step is as much as 6 m.

The step

The upper 5 m of the small cliff are covered with a continuation of the lush coral growth seen at the break in slope above. Overlapping plates of *M. annularis* and *Mycetophyllia* spp. project out from the face of the cliff at about 20–30° like so many overlapping eaves. The amount of coral cover decreases noticeably below about 25 m and toward the base of the step corals cover no more than a fifth of the surface. Correspondingly, the rock surface, covered with coralline algae and a lush growth of gorgonians, alcyonarians, large sponges and *Halimeda*, is more noticeable towards the bottom of the step. Winding down the slope between hummocks that are encrusted by corals and algae are irregular chutes, the continuation of shallow grooves, here also floored with coral rubble and skeletal sand.

The sand slope

The wide, generally featureless sand plains between a depth of 37 and 45 m along the barrier reef and lee side of Glovers Reef are almost identical. The smooth slope is interrupted only by gentle undulating swells and troughs 30–45 m wide, oriented downslope, with the troughs often only discernible by the accumulations of coral sticks, conch shells, blades of *Thalassia* grass and rhodolites along their axes. The surface of fine-grained sand is dotted with numerous small sediment volcanoes, 10 cm or so high, between which clusters of *Penicillus* as well as bushes and clumps of *Halimeda* sprout up here and there, reaching 3–5 plants per m^2 in places. Colonies of *Halimeda tuna* and *Halimeda copiosa* are also seen attached to pieces of *A. cervicornis* rubble in piles brought together by goatfish. These pieces of rubble, together with other fragments scattered about the surface, cover only about a third of the slope and provide a substrate for sponge and octocoral growth.

The surface of the sand slope off South Water Cay (Site 2) (Fig. 3-4), although at a slightly shallower depth of 33 m, and more or less horizontal, is nevertheless identical to the sand slopes described above. Digging with the claw of the submersible confirms that the sediment is also fine sand, to a depth of at least 10 cm.

On the lee side of the atoll (Sites 5 and 6) (Fig. 3-5) almost one-half of the slope is occupied by blocks or mounds of reef rock, generally 1–1·5 m high and 2–3 m long, with occasional large masses up to 6 m high and over 15 m long seen off the eastern, leeward site (Site 5). These blocks are veneered with sporadic growth of corals, gorgonians and alcyonarians.

The slope at the base of the step between 37 and 45 m off Queen Cays is very different from that described above; it has a steeply sloping rock surface (45–55°) veneered with sand and coral rubble and punctuated by knobs up to 3 m high that are either lodged talus blocks or mounds of dead corals, both of which are now crowned and veneered with *Montastraea annularis* and *Agaricia* spp.

Contact between sand slope and brow

Like the break in slope at the top of the step at 20 m, the change in slope at 45 m is covered with lush coral growth, with living coral increasing abruptly to cover more than half the bottom. *Mycetophyllia reesii* appears abruptly at this depth and sheet-like colonies of *Montastraea annularis* as high as 1 m and as wide as 2 m were seen.

The sand slope off South Water Cay is bounded by a steep-sided coral ridge (Fig. 3-3). The leeward side of this ridge is a steep, coral covered slope of 50–60°. The most common coral is rounded to plate-like colonies of *M. annularis*, which together with other minor forms cover about a quarter of the surface near the base and half near the top. Between the plates and rounded coral knobs are many tubular demosponges and an unidentified brown algae. Most corals towards the top of the pinnacle are covered with fuzzy brown algae or gorgonians.

The flat crest of the ridge at 15 m is 20–30 m wide and is covered with a thicket of *A. cervicornis* and intervening small heads of *M. annularis, M. cavernosa* and *Diploria* sp. The steep seaward margin of the ridge is a concave surface dipping from 45 to 55° down to the top of the wall at 65 m. The steeper, upper part of this slope is about a third coral-covered.

The brow

Barrier Reef and lee side of Glovers Reef

Irrespective of what is above, the bottom between about 40 m and the top of the wall at 65 m is similar at all localities; it dips at 45–50° with many irregular ledges, promontories and hummocks, some of which support living coral, and between which are irregular streams of sand coursing down the slope. Living coral covers less than a quarter of the total surface, with overlapping plates of *M. annularis* and *Agaricia* spp. forming individual, isolated mounds or knobs 2–3 m across (Fig. 3-8A) separated by irregular depressions floored with *Halimeda*-rich sand. The growth of plate-like corals is most luxuriant on the lee side of Glovers Reef where colonies cover as much as half the surface. The knobs, hummocks and irregular ledges that do not support living coral are veneered with coralline algae and have lush growths of alcyonarians, gorgonians and sponges (Fig. 3-8B).

Halimeda is conspicuous in this sector of the reef margin growing on irregular rocky knobs in and around coral colonies (Fig. 3-8C) and covering as much as a

Fig. 3-8. Tobacco Cay, site 1, (A) depth 67 m: a small mound, about 4 m wide and 2 m high just above the top of the wall on the reef brow, crowned with plates of *Agaricia grahamae*; (B) depth 60 m: crustose coralline algal encrusted surface between coral colonies veneered with sediment and supporting growth of oscular sponges (arrow) that are *ca* 10 cm long.

quarter of the rock and coral surfaces. Individual colonies of *H. copiosa* and *H. discoidea* more than 0·5 m across, growing as bushes, veils and vines are common. The most abundant *Halimeda* growth at this depth was observed off Queen Cays, where every solid substrate not buried by sand supports *Halimeda*.

Along the barrier reef the depressions between hummocks or mounds are interconnected and floored with sand. By these circuitous routes sand wanders down the slope in a series of irregular, often discontinuous streams, which at any one time cover about half the surface.

In contrast to these meandering streams of sand the ledges on the lee side of Glovers Reef are cut by many straight to sinuous grooves 5–7 m wide and as much as 50 m apart, and oriented downslope. These grooves are floored with blocks, rubble and *Halimeda*-rich sand and there is as much as 5 m of relief between the floor of the grooves and the top of the intervening bosses of coral. The bosses or promontories between grooves are cavernous and often overhanging.

The transition from brow to wall is gradual, with the roll-over between 60 and 67 m. Approaching the top of the cliff, the relief on the slope increases; the difference between hummocks and depressions is commonly more than a metre and there are local promontories resembling isolated blocks capped with living coral that are often more than 6 m high. Ledges and caves become pronounced, increasing for example, from 0·5 m wide at 60 to 1 or 1·5 m wide at 67 m off Tobacco Cay while off Queen Cays ledges up to 6 m long and caves as deep as 3 m are not uncommon at the top of the wall. As the slope and relief increase so the amount of coral cover on hummocks and ledges increases markedly until a third or a half of the surface is covered with many large isolated plate-like colonies resembling giant lily pads (Fig. 3-9B). The most obvious coral is *Agaricia grahamae*, forming plates up to 5 m long and 1·5 m wide and often draping down over the ledge on which it grows (Fig. 3-9A). Plates of *Montastraea cavernosa* are almost as common as *Agaricia grahamae* and grow in a similar fashion as plates up to 1 m across. Scattered amongst these platelike forms are many colonies of *M. annularis* and *Madracis* sp. As above, on the gentler part of the brow, the rock surface is veneered with coralline algae, yellow cylindrical sponges are profuse, *Halimeda* is prolific and all non-living horizontal and sub-horizontal surfaces are covered with sand that streams over the edge in irregularly spaced narrow chutes (Fig. 3-9C).

Glovers Reef, east side

The steep (65–70°) slope below the spur and groove displays the same ridge and furrow structure as in shallow water above (Fig. 3-10A). At the break in slope the relief between crest and trough is as much as 5 m, but decreases progressively with increasing depth to virtually nothing at 40 m (Fig. 3-10B). The tops of the ridges have a coral cover of about 40%, mostly large, flattened overlapping plate-like colonies of *M. annularis* and *D. strigosa.*, as wide as 3 m (Fig. 3-10A). Other common corals are plate-like forms of *Agaricia* spp., *Agaricia lamarcki, Agaricia grahamae, Colpophyllia natans,* and *Mycetophyllia reesii,* and small columnar-shaped *Montastraea cavernosa*. The sides of the ridges are veneered with overlapping sheets of *M. annularis*. These and the plate-like forms on the ridge crest project out into the water from the steep slope like a series of visors or eaves. The rock between living corals is encrusted with coralline algae and has a prolific growth of massive and tubular sponges and shrubs and vines of *Halimeda*. *Halimeda* covers about a third of the surface area with algae growing out from between coral plates and up from the surface encrusted by coralline algae. Sand in the straight to sinuous channels is rarely rippled, but often burrowed and veneered with algal scum, an indication of temporary stabilization.

On the more gentle slope below 40 m (40–50°) to the top of the wall, the ridge and furrow structure is represented by narrow, sinuous streams of sand with intervening

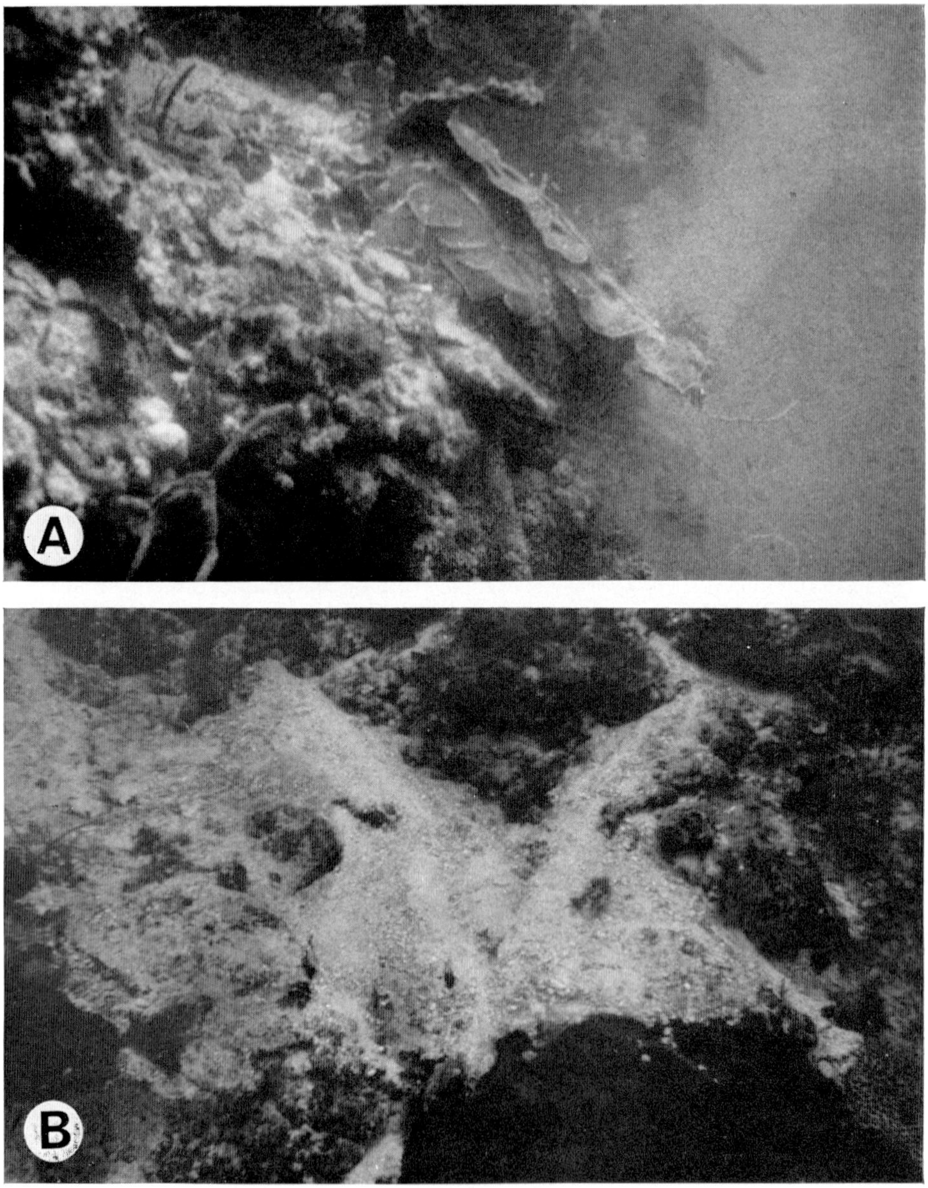

Fig. 3-9. Tobacco Cay, site 1, (A) depth 73 m: the deeper reaches of coral growth with a large colony of *Agaricia*, about 2 m across, plastered on a seaward sloping ledge and growing outward like a visor; (B) depth 70 m: two sand streams coming together to form a cone of sediment at the top of the wall which then spills over a ledge; width of sand stream is about 2 m.

areas of reef rock encrusted by coral, with less than a metre of relief between the two. The crustose coralline surface has less than 15% coral growth, for the most part large plates of *Montastraea cavernosa* and large 'lily pads' of *Agaricia grahamae*. Other common corals are *Agaricia* spp., *Mycetophyllia reesii, Meandrina meandrites, Porites furcata,* and some branching *Madracis* sp. Colonies of *Montastraea* do not

Fig. 3-10. Glovers Reef, site 7, (A) depth 27 m: step mantled by many plate-like and overlapping colonies of *Montastraea annularis* between which are growing oscular demosponges and string-like octocorals. This cliff is cut by many channels, one of which separates this coral growth from that in the background (right); (B) depth 40 m: a 2 m wide groove in the upper part of the brow; filled with sand and scattered coral rubble.

appear to grow below a depth of 62 m. Numerous small *Halimeda* plants, sponges, and alcyonarians are found on the encrusted rock between corals. The narrow grooves are filled with *Halimeda* sand. Occasional large blocks are encountered on this section of the slope, perched precariously on the slope above the wall.

THE WALL

Morphology

From a distance, in the dim light, the surface of the wall looks like the surface of a well stratified outcrop. As one approaches the face, however, the ill defined strata materialize as a series of many irregular, often discontinuous, ledges (Fig. 3-11A), mantled with fine sediment, that are separated by bare rock or small caves and broken by many fissures. It is the variations and combinations of these elements, sediment-covered ledges, even rock surfaces, caves and fissures, that produce variations in the surface topography of the wall.

The upper 30 m of the wall, between the break in slope and a depth of about 90 m, are characterized by a series of small angular ledges, 0-5–1·5 m wide and spaced 2–3 m apart, which can rarely be traced laterally more than a few metres (Fig. 3-11A). Individual ledges are rarely of uniform width, ranging from a few centimetres to a few metres wide along their length. Although the surface of some ledges is subhorizontal, many are inclined seaward, when seen straight on, at angles as steep as 30°. Many of the ledges join and bifurcate with other ledges several times when traced laterally. The walls between the ledges are either bare rock or small caves.

On the lee side of Glovers Reef the ledges are large and the spacing between them may be as much as 10 m. Perched on these ledges are numerous blocks and piles of coral rubble with some blocks as large as 20 m high. The rock between ledges is delicately sculptured with many sharp projections and irregularities, and the ledges are dissected by many vertical fissures and caves and canyons giving the wall a very rugged topography.

In contrast, the upper part of the wall off the eastern side of Glovers Reef is relatively smooth, with subdued relief (Fig. 3-11B). Ledges are normally less than one metre wide, can never be traced laterally along the wall for more than several metres, and are separated by as much as 2 m of rock wall. Often scars on the wall attest to the calving or breaking off of segments or whole ledges. In general the wall has a cellular to pock-marked appearance, with many pits or depressions, but none of the features is larger than 1 m.

A zone of particularly large caves, often as deep into the escarpment as 5 m and up to 8 m across, occurs between 100 and 110 m off Tobacco Cay (Site 1) and between 90 and 105 m off South Water Cay (Site 2). Ledges between caves are also large, 1·5–5 m wide and 3–7 m apart. Off Tobacco Cay this zone could be traced laterally for as much as 200 m and several observers recorded a prominent notch at a depth of 116 m.

Although the caves and ledges are the most striking part of the morphology, the surface is made even more irregular by many smaller scale features. These are in the form of innumerable small rounded holes and cavities, delicate sculptured prongs, small terraces or benches, and columns and arches from several centimetres to 0·5 m in size. Such features give the surface a pock-marked appearance as tortuous and complex as that produced by intertidal erosion.

The subhorizontal layered appearance of the wall is broken by vertical to near-vertical fissures. These fissures, generally 1–3 m wide and up to 30 m long, often extend as much as 5 m into the rock face. The bottoms of cracks are usually choked by blocks of rock, with cones of sand-sized sediment extending out from the face for several metres. Large re-entrants or canyons tens of metres wide and extending back into the

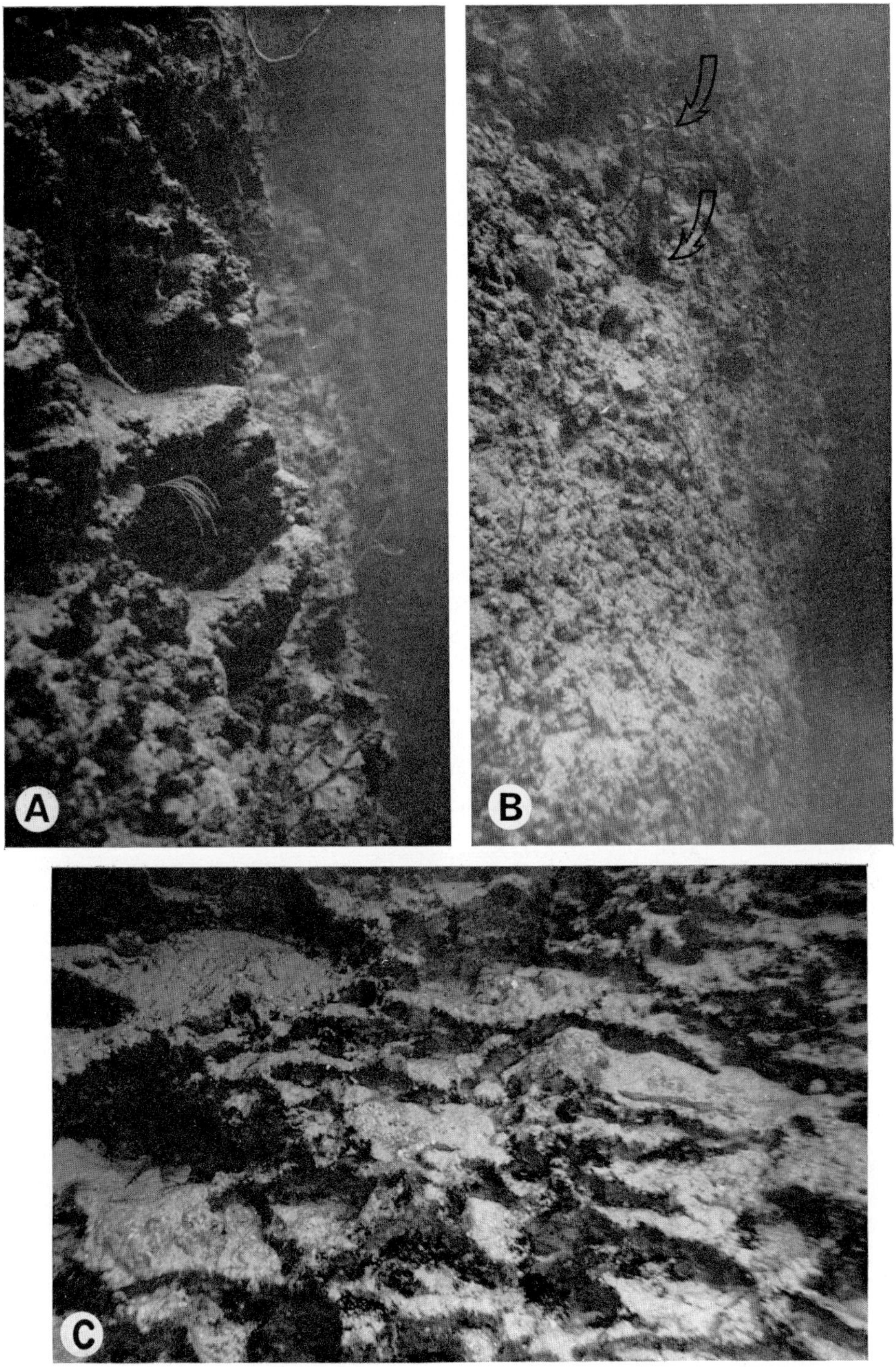

Fig. 3-11. (A) Tobacco Cay, site 1, depth 90 m: a view looking along the wall illustrating the irregular, ledged nature of the surface, the distance between prominent ledges at centre is about 1 m; (B) Glovers Reef, site 7, depth 110 m: a view looking along the wall which is relatively smooth compared to the wall off Tobacco Cay (10A); the gorgonian between arrows is about 1 m long; (C) Tobacco Cay, site 1, depth 100 m; a view looking straight at the wall illustrating the many ledges, each of which is covered with white sediment; ledges are about 0·5 m wide.

cliff face 10–12 m are often encountered when traversing along the wall. Rare, curving, subhorizontal fissures mark the upper limit of large pieces of the wall that have become detached and slumped. In one such slump where the fallen block was still in place, a slab about 100 m wide was observed to have dropped 10–15 m.

On the eastern side of Glovers Reef (Site 7) and Queen Cays (Site 6), the relatively smooth wall becomes very rugged at a depth of 100–105 m with many large ledges, huge caverns and fissures. Off Glovers Reef this change is marked by a huge ledge that extends out from the wall 8–10 m at a depth of 103 m. Off Queen Cays the slope is made up of steep vertical segments with many ledges and caves with intervening more gently dipping areas of perched sediment and talus. Inclined talus slopes are rarely more than 20 m high and as they form the top of large caverns, it is difficult to differentiate this topography from just a steep scalloped surface. Off Glovers Reef the caves are smaller and individual ledges are rarely longer than 5 m. Particularly conspicuous at a depth of 116 m is a zone of prominent ledges. Vertical walls between caves and ledges are very iregular; yet the surface is often very smooth (Fig. 3-12B). Characteristic of this zone, both off Glovers Reef and off Queen Cays are numerous large blocks, one of which was measured as more than 20 m high and 30 m long, perched precariously on overhangs and ledges.

The base of the wall is not straight in plan, but composed of more or less equally spaced convex protuberances or rock bosses that extend out some 5–8 m, are 10 m or so wide and separated by 10–15 m wide re-entrants. Wedges of sediment and talus are commonly piled up in the intervening depressions.

Sediments of the wall

The horizontal and subhorizontal elements of the wall are enhanced by accumulations of white sand and mud on all near-horizontal surfaces (Fig. 3–11C). This

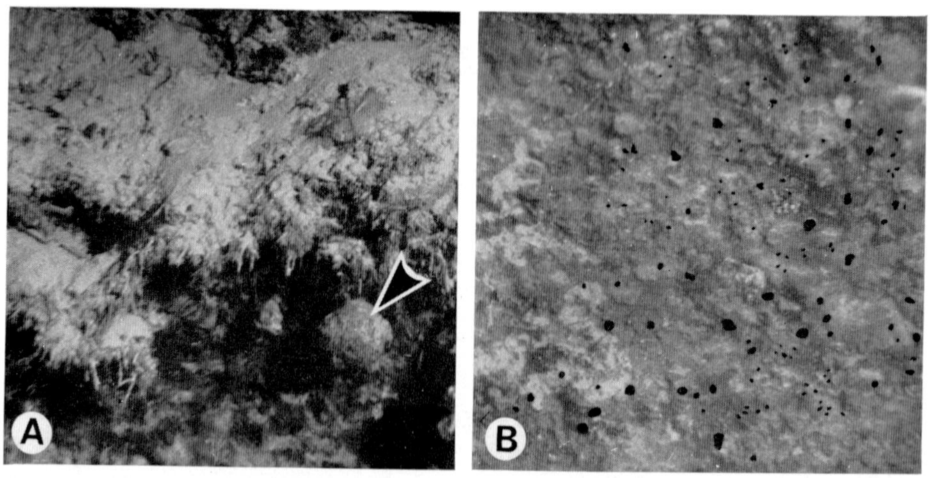

Fig. 3-12. Glovers Reef, site 5, (A) depth 110 m: a close-up view of a small ledge, about 30 cm wide the surface of which is mantled with mud and occasional plates of *Halimeda*; note the organic filaments extending down from the edge of the ledge and sclerosponge growing below the ledge (arrow); (B) depth 110 m: a view looking at the rock wall between two ledges. The light splotches are encrusting sponges and the many holes (dark spots outlined in ink) are the papillae of boring sponges, photograph 0·2 m across.

sediment funnels down from the brow above and is produced locally on the reef at the break in slope at 65 m. Streams of sand that wind their way over the top of the wall produce sediment cascades resembling small waterfalls out of hanging valleys or continue to meander down the face of the cliff along the inclined ledges, like a series of switchbacks in the mountains. Thus ledges collect a surprising amount of sediment (Figs 3-11C, 3-12A); the upper surface of almost every ledge has a wedge that may be up to 10 cm thick at the back of the ledge. The sediment is commonly observed to be medium to coarse-grained sand, principally *Halimeda* plates, mixed with mud (Fig. 3-12A). Sediment also moves down in suspension. During the dives we saw numerous falls of sediment that show how differential settling separates the different size fractions; the finer equant grains settle faster than the larger oatmeal-like *Halimeda* plates, which flutter slowly down.

Not all the sediment is sand and mud; towards the base of the wall, ledges have numerous plates of coral mixed with *Halimeda*-rich sand. Even caves are floored with sand and mud and we saw *Halimeda*-rich sand on the floor of one cave as much as 3 m in from the entrance, an indication of the cavernous porosity of the wall.

Benthic organisms of the wall

The wall is biologically a transition zone, the graduation from the shallow-water reef community dominated by hermatypic corals, octocorals, green algae and crustose coralline algae to a deeper water community of coralline algae, octocorals, ahermatypic corals, demosponges, sclerosponges and endolithic sponges.

The upper 20 m or so of the wall have the deepest living representatives of the shallow-water reef above (Fig. 3-13). Hermatypic corals, although not prolific, are relatively common on the rock surface to a depth of 70–80 m. Below 80 m, however, the abundance of even these deep water forms decreases markedly and the deepest-growing hermatypic corals are seen at 93 m along the barrier reef and at 102 m on the eastern side of Glovers Reef.

The most obvious corals growing between 67 and 80 m are *Agaricia grahamae* (up to 3 m in diameter), *Agaricia fragilis, Montastraea cavernosa, Madracis* sp., *Solenastrea* sp., *Stephanocenia* sp., and *Mycetophyllia reesii*. These corals veneer small ledges and protuberances on the rock face and together cover only about 10% of the surface area. Those colonies that are plastered on vertical walls are slightly larger, up to 15 cm in diameter, than those growing on subhorizontal surfaces which are rarely more than 10 cm across. Particularly conspicuous are *Agaricia* colonies growing on slightly inclined ledges and projecting out from the wall (Fig. 3-9A). Clusters of *Agaricia* growth may contain as many as 18 colonies per m^2. Colonies of *Madracis* sp. growing between depths of 67 and 71 m are small, rarely larger than 15 cm in diameter. Between 80 and 102 m the only large, living hermatypic corals seen were *Madracis* sp. and *Agaricia fragilis*.

Clusters of *A. fragilis* containing up to 10 colonies per m^2, with individuals as large as 10 cm, grow on small protuberances and ledges facing seaward, to a depth of 90 m. The deepest, living, attached corals seen along the barrier reef were small *Madracis* sp., 3 cm in diameter, growing in greenish algal turf at 93 m off Tobacco Cay, while the deepest off Glovers Reef was a small *A. fragilis* at 102 m.

At 95 m off Tobacco Cay an unattached flattened colony of *Montastraea cavernosa* 20 cm long and 10 cm wide, with brown colour, suggesting it was alive, was seen precariously perched on a small ledge.

The absolute depth limit of hermatypic corals varies from locality to locality (Fig. 3-13) and is a function of both the area and the intensity of observation. In general, however, the above descriptions apply to all localities, with the zone of abundant growth terminating between 70 and 80 m and the deepest living hermatypes extending to 100 m.

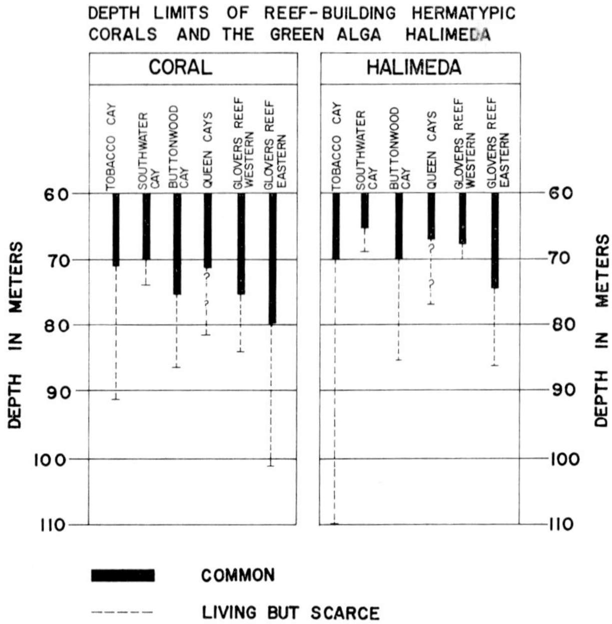

Fig. 3-13. A diagram illustrating the depth limits of the green alga *Halimeda* and reef building corals at each of the dive sites, as observed from the submersible.

Numerous shrubs, veils and vines of *Halimeda* grow in between and around the plate-like colonies of hermatypic corals. These growths commonly drape down over the ledged surface in profusion to a depth of 70 m and in places 76 m. *Halimeda* decreases in abundance abruptly and disappears completely at many localities below about 75 m. The deepest living *Halimeda* (cryptica?) was seen growing at a depth of 110 m off Tobacco Cay (Fig. 3-13).

The corals and algae described above are really the deepest stragglers of the lush shallow reef community between 70 m or so and the surface. Most of the wall is veneered with a community of encrusting and cryptic fauna. The irregular topography of the wall has created three slightly different habitats: (1) the vertical rock surfaces, (2) horizontal ledge surfaces and (3) caves and the undersides of ledges.

The vertical rock surfaces, which from a distance appear brownish in artificial light, are partially veneered with encrusting organisms. The most conspicuous element of this veneer is crustose coralline algae, light pink and dark red in colour, that on the upper part of the wall covers more than half of the exposed surface. The second commonest organism encrusting the rock face is a yellow sponge, which although small (less than 10 cm in diameter) covers an estimated 15% of the surface (Fig. 3-12B). These sponges are most common in the shadow of overhangs, where crustose coralline algae are less abundant. The wall is also veneered with a green fuzzy growth of what is thought to be green algae. The only upright organisms growing on the rock wall are alcyonarians, which are attached to the wall and grow out at angles to it.

Even though more than half of the ledges are covered with sediment, some are free of sediment or protected from sediment fallout by larger ledges. The sediment-free ledges, both between corals near the top and on the main part of the wall below, are invariably covered with crustose coralline algae. This cover is most prolific between 75 and 90 m where scab-like encrustations 15–20 cm wide cover 30–50% of all exposed surfaces. These crusts or protuberances decrease in size and abundance with depth, and at 110 m, less than 20% of the surface is covered with colonies, often only a few centimetres across. On most ledges colourful yellow, orange and blue encrusting sponges grow between the algae. In addition, erect demosponges and octocorals project up from many surfaces.

The undersides of ledges and interiors of caves and caverns are often veneered with coralline algae but the most conspicuous organisms are tubular sponges (*Agelas*), gorgonians hanging down into cavities, and forests of small white ahermatypic corals growing on the roof.

Sclerosponges (Fig. 3-12A) occur on the tops of ledges protected from sediment fallout, on the undersides of ledges and in cavities. *Certoporella nicholsoni* (3–20 cm in diameter) is the most common sclerosponge in this cryptic habitat, reaching a density of 1 per m^2 in some cavities at a depth of 100 m. *Goreauiella auriculata,* although less common, was seen growing at similar depths. Sclerosponges appeared to be more abundant at the leeward, western side of Glovers Reef (Site No. 5) than at any other. They were most commonly seen beneath ledges and in caves between 70 and 105 m where they are large (50 cm) orange specimens of *Ceratoporella*. Sclerosponges were seen in cryptic habitats to the bottom of the wall and where it passes into the cliffed fore-reef as off the Queen Cays (Site 4). They were noted commonly to a depth of 130 m, only rarely deeper, and not at all below about 150 m; this depth (150 m) may not be their absolute depth limit but rather the lowest depth to which they commonly occur.

THE SLOPING FORE-REEF

Introduction

If sea level were lowered 500 m or more, the sloping fore-reef around the margins of the trough between the barrier reef and Glovers Reef would look like a series of coalescing alluvial fans lapping up on the base of an escarpment 60 m high. The following sections summarize our observations on the morphology, sediments and organisms of this surface, and are based on the profiles along the barrier reef (Sites 1, 2, and 3). The fore-reef on the leeward side of Glovers Reef (Sites 5 and 6), while basically the same as along the barrier, also has some major differences, which are outlined in a following section.

All sites are roughly similar in their morphology and consist of three zones: (1) a slope of blocks, coral debris and skeletal sand that is steep adjacent to the wall and flattens out basinward to a depth of about 200 m, which we call the *proximal* sloping fore-reef; (2) a very gentle slope of muddy sand with few obvious corals or blocks that continues to decrease basinward to a depth of 300 m and more, which we call the *distal* sloping fore-reef; and (3) the flat floored trough composed of pelagic carbonate sands and muds.

The proximal sloping fore-reef along the barrier reef

Morphology and sediments

From a distance the upper part of the sloping fore-reef resembles a scree slope; a jumble of talus blocks and rubble footing a steep cliff extends out and down into the gloom as a relatively smooth sediment slope punctuated by occasional stranded blocks.

The contact between the wall and slope is formed by cones of rubble lapping up against the wall between rock bosses (Fig. 3-14A). The rock bosses (Fig. 3-14C), each of which is 12–15 m wide at the base of the wall, dip seaward from 70 to 90°, and are not as steeply inclined as the wall above. The talus cones between, up to 6 m high and from 7 to 10 m wide, dip seaward at 40–50° (Fig. 3-14B). Because of these irregular talus cones and intervening rock bosses, as well as slumped blocks piled up against the base of the wall, the depth of the contact between wall and fore-reef varies: 112–115 m off Tobacco Cay (Site 1), 110 m off South Water Cay (Site 2) and 110–106 m off Buttonwood Cay (Site 3).

The upper part of the sloping fore-reef adjacent to the wall is very irregular; the slope is often between 35 and 40° (Fig. 3-14A). The bottom is a jumble of blocks, coral plates and fragments, and coarse *Halimeda*-rich sand; near the wall irregular blocks and coral fragments cover almost half of the bottom (Fig 3-15A). Individual blocks and boulders range from 1 to 10 m high (Fig. 3-15B); they are clearly derived from the wall, since blocks in all stages of separation were seen, from slumped slabs of the wall, to pieces that have fallen away, to masses separated from the wall by many metres. Coral fragments are whole and fractured plates up to a metre or more across, but only a few centimetres thick, characteristic of the platey colonies that are seen at the break in slope above (Fig. 3-16C). Particularly conspicuous amongst the coral and limestone debris are *Strombus* sp. shells, with as many as 2–3 per m^2 scattered over the sediment surface. Because this Queen conch is used for bait by local fishermen, these concentrations may not be natural, although fractured shells were not conspicuous.

The sand resembles dried oatmeal because it is composed predominantly of *Halimeda* plates, mixed with fine skeletal sand and silt (Fig. 3-16A, C). The sediment surface is streaked, suggesting downslope movement, and streams of sand with little coarse debris wind down the slope for short distances between the coral plates and limestone blocks (Fig. 3-15A). The presence of streams and cones of talus dammed up behind many blocks and behind coral plates, along with the relatively heavy growth of epibionts on blocks (Fig. 3-16B) and some plates indicate that much of the material has probably not moved in the last several months or possibly years.

The zone characterized by numerous blocks varies in width from site to site but is generally narrow: off Tobacco Cay (Site 1) the zone extends to a depth of 150 m, off South Water Cay (Site 2) to 130 m, and off Buttonwood Cay (Site 3) the zone of blocks is absent and only scattered small limestone blocks, less than 2 m in size are seen at the base of the wall.

Below the zone of abundant talus blocks, whatever the depth, the majority of the fore-reef is an even sand surface (Fig. 3-17A), becoming progressively more gentle

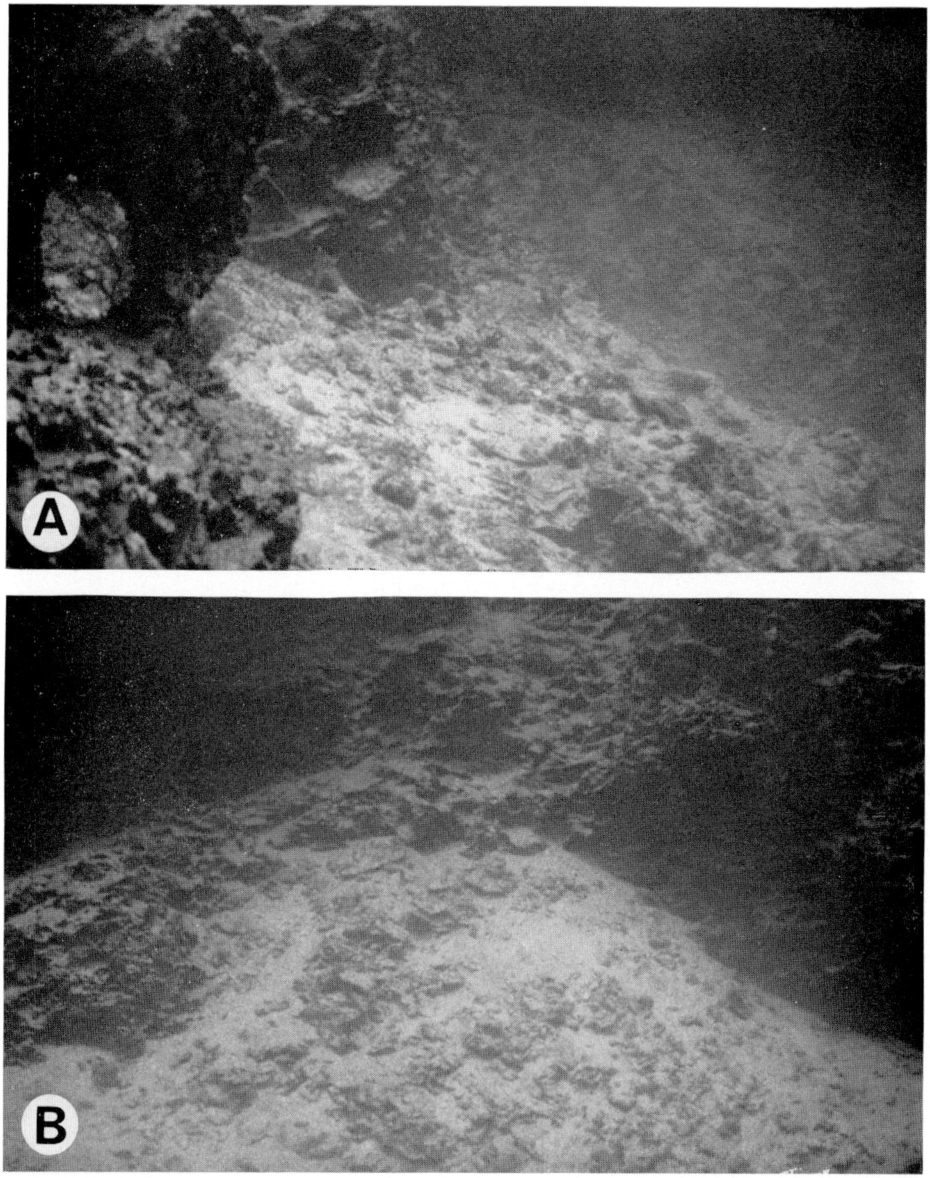

Fig. 3-14. Tobacco Cay, site 1, (A) depth 116 m: a view looking along the contact between the wall and fore-reef from one promontory to another illustrating the irregular surface of the promontories. The cavernous nature of the wall is illustrated by the small hole in the wall at centre left about 2 m high. The steeply dipping proximal fore-reef slope is composed of blocks, coral plates and *Halimeda*-rich sand; (B) depth 115 m: facing the contact between the wall and the fore-reef slope, illustrating a cone of talus that is about 15 m wide at the base and piled up in a re-entrant.

basinward. The slope composed of *Halimeda*-rich sand resembles an open slope of powder snow on a winter morning, with downslope streaks and lobes of white sand on a background of greyish sand, like new snow covering yesterday's fall. In some regions these light and dark sand areas give the impression of recently moved and

Fig. 3-14C. Tobacco Cay depth 110 m: looking obliquely along the base of the wall in an area where the promontories are particularly well developed, the distance between rock bosses is about 20 m.

stabilized sand respectively. In other areas the surface of the smooth slope is made up of undulating, gentle ridges and swales, with less than 1 m of relief and an amplitude of 3 m or more, trending downslope.

The sandy slope is littered with numerous *Strombus* shells and green blades of turtle grass (*Thalassia testudinum*). Off South Water Cay (Site 2), and Buttonwood Cay (Site 3), small boulders and coral plates are scattered on the surface and occasional piles of boulders and coral plates were found down to 150 m, but these coarse-grained sediments cover no more than a fifth of the surface. Off Tobacco Cay (Site 1), the whole slope is punctuated by many scattered talus blocks as well as smaller rocks, protruding out of the sediment. These rocks could not be moved by nudging them with the submersible, suggesting they are monadnock-like exposures of bedrock or huge blocks.

The dip of the sloping fore-reef in this region grades slowly from 35° near the wall to about 25° at a depth of 180 m. Although the dip is generally consistent at similar depths at different localities, the composition of the surface sediment is not. Off Tobacco Cay (Site 1), and South Water Cay (Site 2), the *Halimeda*-rich sand grades into fine-grained sand with only scattered *Halimeda* plates at about 180 m. Digging into this fine sediment at about 190 m off Southwater Cay, however, indicates that at one locality at least, the fine, often muddy sand overlies *Halimeda*-rich sand. Off Buttonwood Cay the coarse *Halimeda* sand occurs to depths of 220 m. The sediment surface in these deeper reaches is often bioturbated with numerous cones (Fig. 3-17A) 5–10 cm high dotting the surface and reaching an abundance of 1 per m^2.

The only physical sedimentary structures seen on the lower parts of the proximal sloping fore-reef are small slump scars and scour pits. The scars indicate downslope sediment movement that occurs around talus blocks and *Strombus* shells to depths of 200 m.

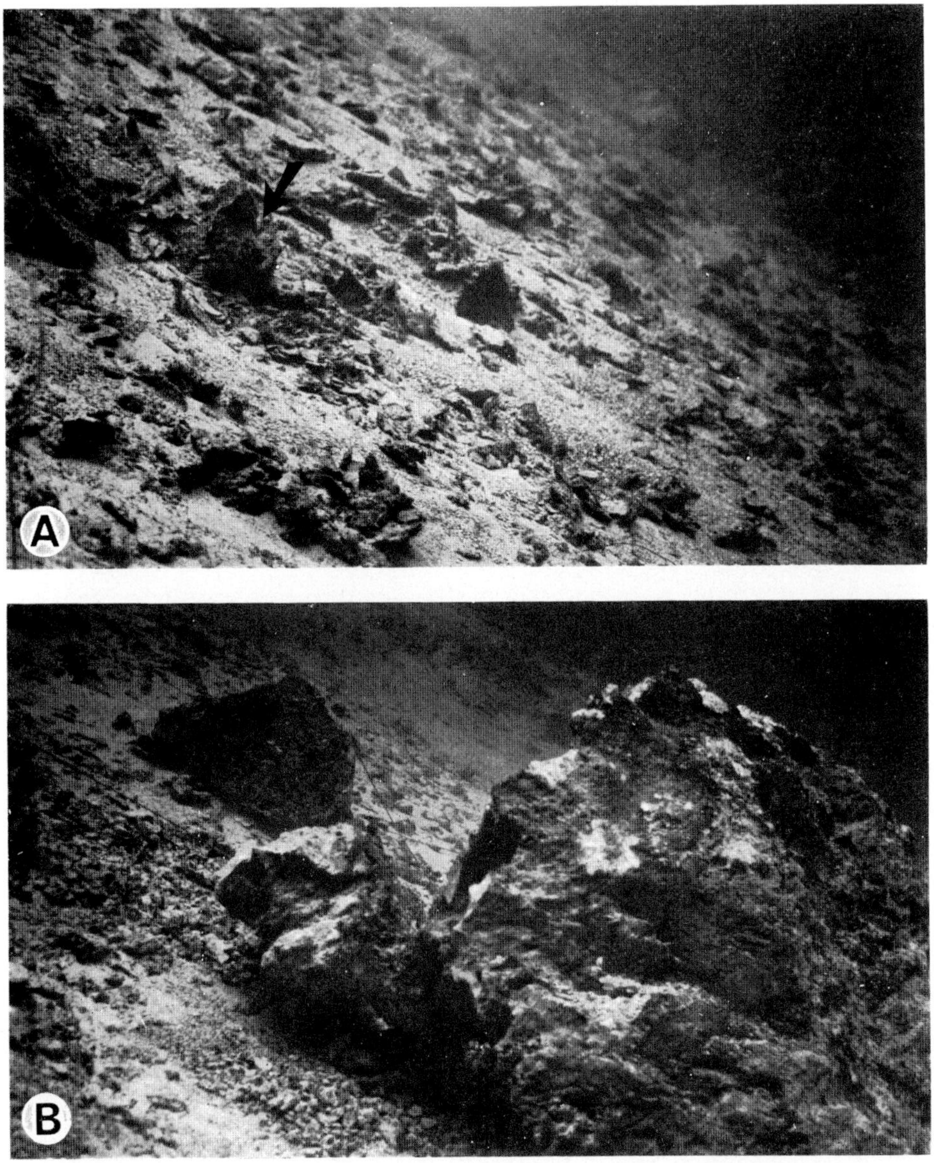

Fig. 3-15. Tobacco Cay, site 1, (A) depth 130 m: looking across the steeply dipping proximal fore-reef slope composed of coral plates and blocks of limestone surrounded by *Halimeda*-rich sand, streaming downslope; the small block in the centre (arrow) is about 1 m high; (B) depth 130 m: close view of a limestone block on the fore-reef slope, some 5 m high, the top of which was blown off during sampling.

Organisms

Most large limestone blocks and coral plates are thoroughly encrusted with epizoans (Fig. 3-16B). Massive yellow, orange and white sponges are the most abundant forms. Antipatharians, *Caryophyllia* (a large solitary ahermatypic coral), ramose pink and white or white ahermatypic corals, red crinoids, serpulid worms and the

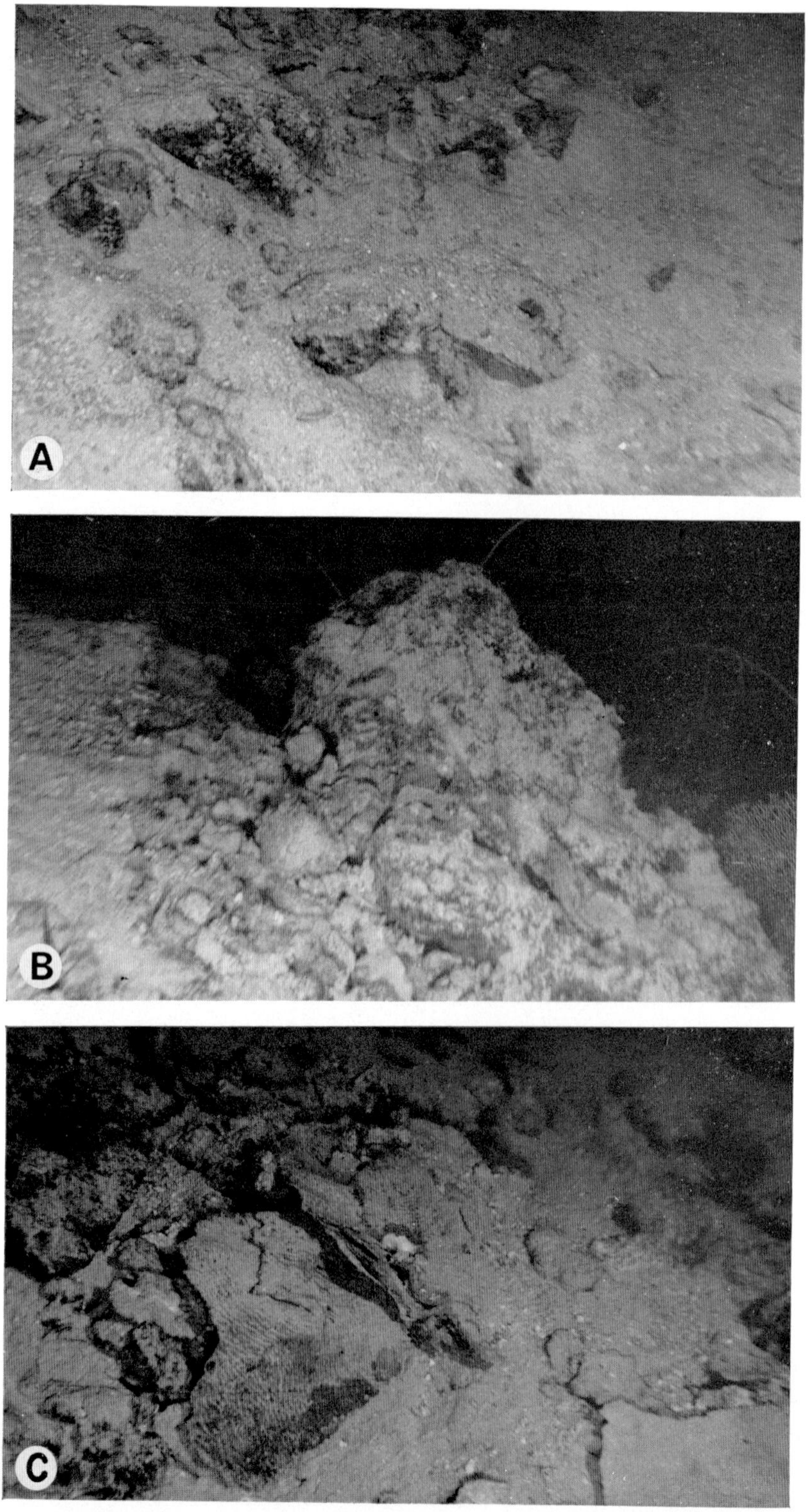
A
B
C

foraminifera *Homotrema rubrum* are common. The sclerosponges *Goreauiella auriculata* and *Ceratoporella nicholsoni* were observed in crannies and under small rocks to a depth of 130 m. Crinoids are commonly attached to the rocks and often emerge from the sediment surface, suggesting rock beneath.

The action of benthic animals is especially noticeable on and in the finer sediment covering the lower parts of the sloping fore-reef where the surface is textured with their tracks and trails. Most tracks and trails appear to be caused by crustaceans (2–5 cm long), hermit crabs, and striped burrowing fish. These fish are commonly observed lurking near the entrances of rubble-lined burrows. Close examination of the sediment reveals an abundance of small benthic fish and invertebrates churning up the upper few centimetres of the sediment.

The proximal sloping fore-reef on the leeward side of Glovers Reef

The configuration of the proximal sloping fore-reef at sites 5 and 6 (Fig. 3-4) on the western, leeward side of Glovers Reef is similar to the sloping fore-reef some 25 km away on the other side of the trough along the barrier reef, yet it has some significant differences, which are listed below.

(1) The blocks at the base of the wall are more numerous and larger; they are up to 15 m high and as wide as 18 m at the base.

(2) The sloping surface (45° decreasing to 35°) between blocks varies from a hard rock floor, occasionally veneered by sand, to a series of gently undulating ridges of sediment oriented downslope with rock exposed in the swales. The surface of the rock is ledged, suggesting that it may be cemented talus.

(3) The sediment on the slope appears (upon observation from the submersible) to contain much less *Halimeda* than along the barrier reef at similar depths.

(4) The morphology of the deeper portion of the proximal fore-reef varies from place to place; at site 6 the slope gradually decreases with depth, as along the barrier reef; at site 5, however, between 165 and 185 m there is an enormous outcrop or buried block. In places the rock rises almost vertically as much as 30 m from the slope on the upslope side, is about 10 m wide at the top and then forms a vertical face 50 m high from a depth of 135–185 m on the downslope side. In other places the 'outcrop' is just an irregular rock face from a depth of 168–185 m. Parts of this escarpment have either collapsed or are overridden with sediment. The lateral extent of the feature is not known as it was only observed on the one traverse at this locality that went this deep. The face is very deeply scalloped into small holes or cavities, not as sharp surfaces, but rather as smooth outlines resembling innumerable swallows' nests in a cliff. It is very much like the very deep cliffs at 300 m off the eastern side of the reef (site 7). The surface is cut by many V-shaped gullies and canyons that are often partially filled with sediment and exhibit blocks wedged in the openings. Small continuous, irregular sinuous fissures and cracks, 20–30 cm wide can be traced at times about 30 m up the cliff. The rock surface exhibits a moderate coverage of crustose coralline algae

Fig. 3-16. Tobacco Cay, site 1, (A) depth 120 m: small talus blocks of limestone and some coral surrounded and partially buried by *Halimeda*-rich sand; (B) depth 125 m: a small block, *ca* 3 m high, on the sediment slope and almost buried by sediment. The covering of attached epibionts indicates that it has not moved for some time; (C) depth 119 m: a close view of a jumble of coral plates. The large plate that is about 0·5 m across at left centre is likely *Agricia* sp. The coral plates are covered with *Halimeda*-rich sand.

and a profusion of small blue sponges, yellow sponges, lithistids, and crinoids. A small pile of talus has formed at the base of this cliff, with blocks accumulated at the openings of gullies and canyons and leaning up against the cliff. Sediment in front of the block is finer with only scattered *Halimeda* plates. This sediment is burrowed and exhibits many tracks and trails.

The distal fore-reef

There is a marked decrease in inclination of the sloping fore-reef at depths between 210 and 220 m, from 15 or 20 to 10° or less. This change in slope is coincident with a change in sediment from *Halimeda*-rich sand or sand with scattered *Halimeda* plates to fine sand and silt (Fig. 3-17B) with no recognizable *Halimeda* plates of fragments. The surface of the monotonous slope below 220 m is textured with the tracks and trails of organisms and pock-marked by innumerable burrow mounds up to 5 cm high and reaching a density of 10 per m^2. Like the proximal fore-reef above, the distal part of the fore-reef is littered with conch shells and blades of *Thalassia* grass.

This monotonous, relatively flat sediment slope continues seaward without interruption except on the leeward side of Glovers Reef where isolated outcrops are encountered to depths of at least 280 m, the deepest area visited. Seismic profiles and piston cores suggest that similar fine-grained sediment floors most of the trough between the barrier reef and offshore atolls (see Chapter 2).

THE CLIFFED FORE-REEF

Introduction

The cliffed fore-reef that occurs along the margins of the Cayman Trench at Sites

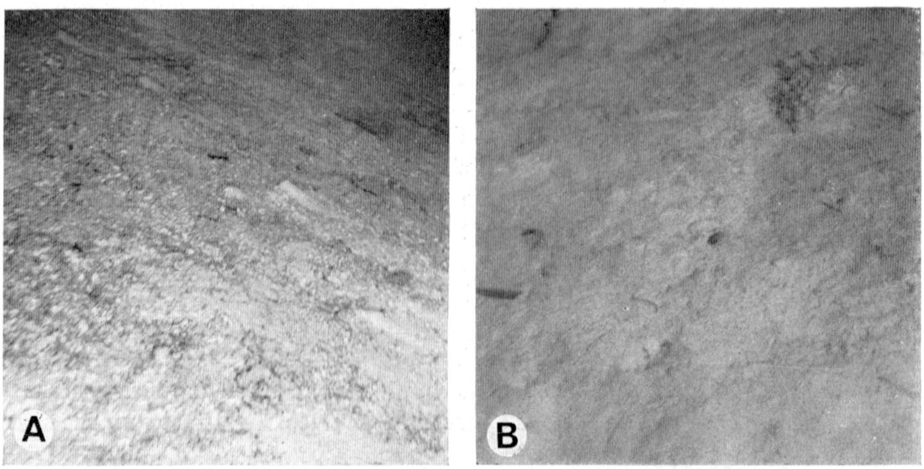

Fig. 3-17. Tobacco Cay, site 1, (A) depth 145 m: a smooth basinward dipping slope of *Halimeda*-rich sand broken by small sediment volcanoes due to the action of burrowing infauna; note the asymmetry of the mounds with downslope tails, resulting in a slow, downslope creep of sediment, each mound is 5–10 cm high; (B) depth 260 m: the smooth surface of the distal fore-reef composed of mud rich in pelagic skeletons and bioturbated by infauna, the field of view is about 1·5 m.

4 and 7 (Fig. 3-6) consists of alternations of steep rocky cliffs and gentle talus slopes. Most of the following description is based on observations at the Glovers Reef (Site No. 7), which is outlined in Fig. 3-6, but features observed in the Queen Cays profile along the barrier reef (Site 4) are also included.

The fore-reef can be informally subdivided into four different segments: (1) the uppermost tred, a talus slope extending from the base of the wall to a depth of about 150 m, (2) a series of sediment treds and small rock walls, cut by a series of down-slope trending ridges and furrows, between about 135 and 210 m (the ridge and furrow zone), (3) a relatively gently dipping sediment slope to 310 m, (4) a steep to overhanging cliff to some unknown depth.

Upper talus slope

The pile of talus that buries the wall in various places (Fig. 3-18A) extends down from 120 m or so to the top of the first of many small cliffs somewhere between 135 and 150 m, depending upon the locality. The blocks of limestone and coral plates appear to be keyed together forming a slope in excess of the angle of repose and are best termed 'perched talus'. The perched talus slope is commonly inclined at 30° or more and consists of many platey coral fragments, particularly *Agaricia* and blocks of limestone (Fig. 3-18B) both of which are covered with coralline algae. In addition, there are blocks and heads of *Diploria* and *Montastraea,* and piles of *Acropora cervicornis*. These large clasts are surrounded by *Halimeda*-rich sand, which comprises at least half of the talus material (Fig. 3-18B). At one locality a large curving 'amphi-theatre' marks a slump tens of metres wide.

Fig. 3-18. Glovers Reef, site 7, depth 140 m; one segment of the steeply dipping cliffed fore-reef composed of a jumble of coral plates (averaging 0·5 m across) along with scattered limestone blocks and *Halimeda*-rich sand forming a slope of about 30°.

Ridge and furrow zone

Between 135 and 150 m the slope is increasingly broken by a series of small cliffs that become a persistent aspect of the bottom to a depth of about 210 m, where they die out. Between the cliffs, steeply dipping rock and sediment slopes are corrugated into a series of ridges and furrows oriented downslope (Fig. 3-19). Ridges stand 3–5 m above the intervening furrows; both ridges and furrows are 5–10 m wide. The inclination of this slope ranges between 45 and 50°.

Ridges and furrows

The rock that makes up the ridges, when viewed from the side, gives the impression of a series of overlapping ledges dipping seaward at 30–50° (Fig. 3-19). The rock is very porous with numerous holes, caves (Fig. 3-19B) and irregular voids, always with smooth rounded profiles. In fact all surfaces are rounded and very few are angular as in the shallow profile on the wall. Many of the holes are bridged by very thin plates of rock. In contrast to the rock forming the wall above, in which nothing can be seen of the internal composition, the rock of the ridges is clearly seen to be composed of many shallow water corals, in all orientations, and there is the suggestion that many of the small rounded ledges are composed of platey corals.

Loose sediment floors the furrows between ridges, forms cones against the base of intervening cliffs (Fig. 3-20A), and blankets the ridges (Figs 3-19B, 3-20B). The sediment is a poorly sorted jumble of boulder-size blocks of rock, pebble and cobble-size fragments and whole plates of *Agaricia* spp. (Fig. 3-20C, D), sticks of *A. cervicornis,* heads of *Diploria* up to 1 m in diameter, *Strombus* shells and *Spondylus* valves, all surrounded by *Halimeda*-rich sand (Fig. 3-20B, C, D). This sediment veneers all horizontal and subhorizontal surfaces like a mantle. In the furrows large coral boulders, talus blocks and plates of coral often form dams behind which cones of rubble up to 10 m long accumulate.

Cliffs

Many of the cliffs (Fig. 3-20A) that alternate with this ridge and swale topography are slump scars because the resulting smooth surface cuts across numerous coral heads and intervening sediment. Others give no clue as to their origin, like the wall higher up on the profile. In this zone the cliffs form as much as half of the slope; segments are from 5 to 20 m high with ledges 1 or 2 m wide separated by caves or sculptured surfaces and only rarely are they vertical and smooth. A series of particularly large caves and promontories occurs at depths of about 165 m (three separate observations).

Each ledge is covered with *Halimeda*-rich sand up to 10 cm thick (Fig. 3-20A) and scattered coral plates that have fallen from above. The ledges are dissected by numerous open fissures that extend vertically for 10 m or more. The bottoms of these fissures are invariably choked with a talus cone of sediment and coral rubble.

Most of the rock surface, whether in cliffs or ridges, is devoid of visible marine organisms. Crustose coralline algae rarely cover more than 10% of the surface. They are most abundant on subhorizontal surfaces oriented toward the dim light. The commonest large animal is a white, button-like *Lithistid* sponge. Other common, though not abundant, epibionts are tunicates, encrusting white and yellow sponges, alcyonarians, anemones and crinoids. Growing upward from the ledges are large

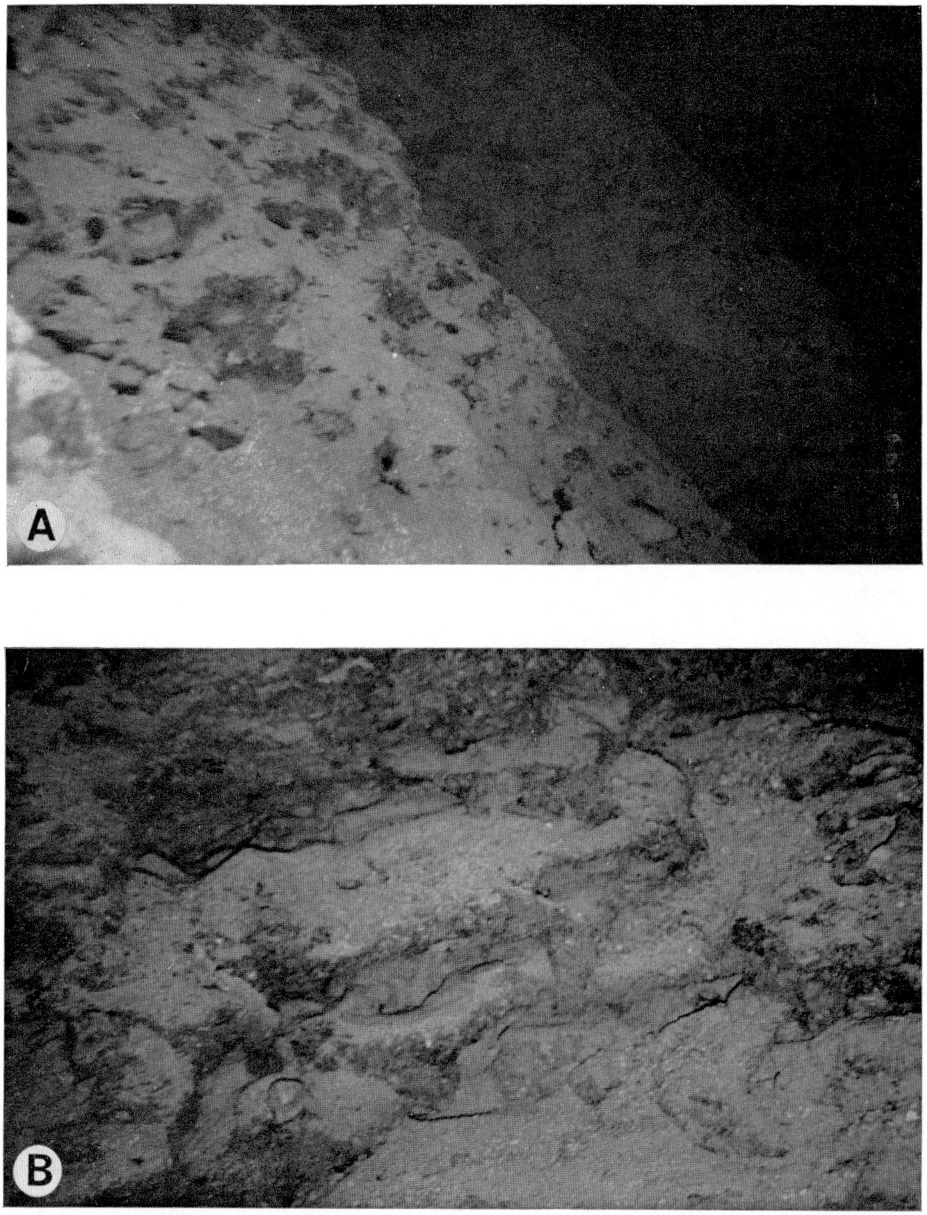

Fig. 3-19. Queen Cays, site 4, (A) depth 170 m: a view across one of several ridges of coral rubble veneered with sediment; in the background about 15 m away is another ridge. The slope of the rubble spur is nearly 50°; (B) depth 170 m: a close view of the face of one of the ridges in part A composed of a layered rock surface, covered with sediment.

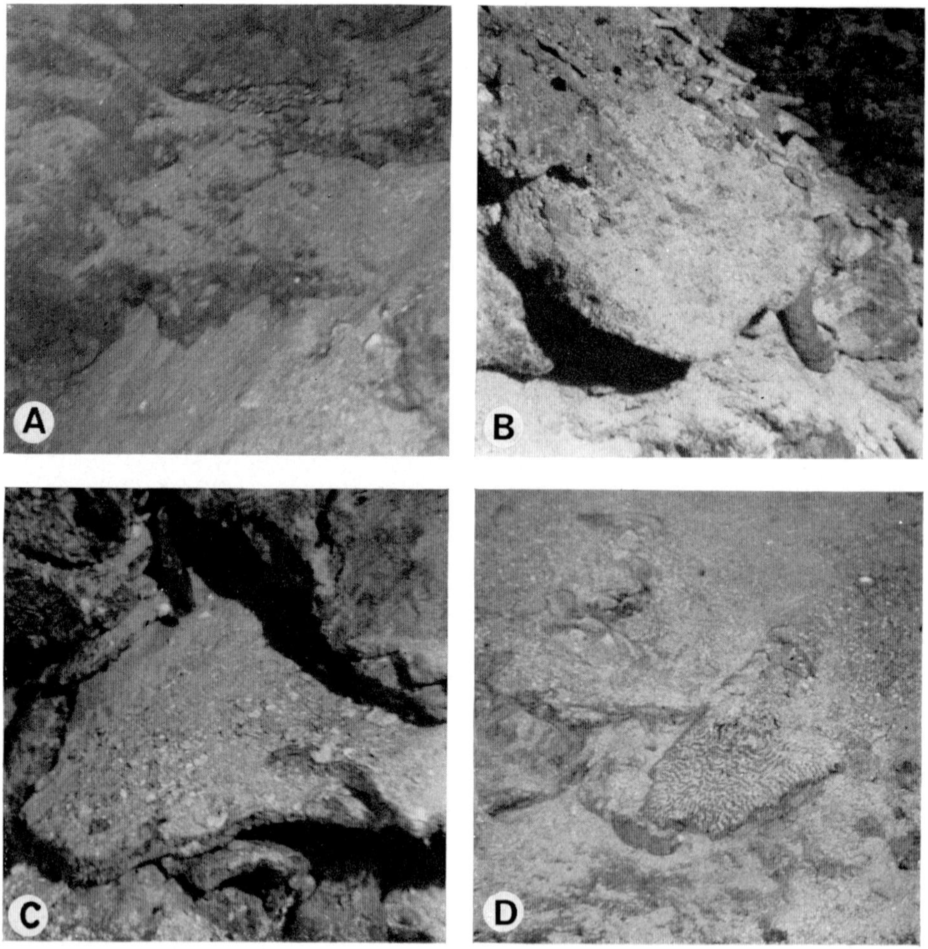

Fig. 3-20. Glovers Reef, site 7, (A) depth 145 m: the base of a small cliff on the cliffed fore-reef against which *Halimeda*-rich sand has piled up, the cliff is about 2 m high; (B) depth 210 m: perched talus and sediment composed of stick corals (top right) and large and irregular coral boulders, with the one pictured here being about 2 m across; (C) depth 200 m: a close view of the coral and sand talus; note the thin coral plate (3 cm thick) veneered with *Halimeda* sand and shingled nature of the plates; (D) depth 183 m: numerous coral plates (the one in the centre is *Mycetophyllia* sp.) perched on a small ledge on one of the rock ridges; the coral plate is about 0·5 m wide.

basket-like sponges, soft corals, gorgonians and antipatharians. Numerous small blue tunicates, clusters of *Thalamophyllia,* ramose white ahermatypic corals, and orange to white stylasterines grow beneath small overhangs.

Sediment veneered rock slope (210–305 m)

At a depth of about 210 m off Glovers Reef and 240 m off Queen Cays the pronounced ridge and furrow structure gradually dies out.

Off Queen Cays we did not dive deeper than 280 m (Fig. 3-6) and in this lower zone the bottom in most places is an even surface of *Halimeda*-rich sand dipping seaward at about 45°, with numerous longitudinal sand ridges or rills with small levees. Coarse *Halimeda*-rich sand is concentrated in the centre of these channels with the finer material on the levees. These appear to be extensions of the ridge and furrow system above, and these swales become more numerous as they bifurcate downslope.

Observations of Glovers Reef, however, continued to 320 m and give a much more comprehensive view of this zone. Below 210 m the bottom is a series of vertical cliffs but the slopes between are more gentle, and often have only a dusting of sediment over rock. Between cliffs the sediment slopes vary from 50° at 210 m to as low as 30° near 300 m. The number of cliffed segments decreases with depth and from 260 to 310 m the bottom is a relatively smooth sediment surface.

The rock surface mantled by sediment in the lower reaches is not a continuous smooth surface but rather a surface similar to the surface of the ridges above, composed of many mini-steps rarely more than half a metre high, that are inclined at 30–40° seaward (Fig. 3-23C). These mini-steps and the inclined surface have many more holes and cavities than does the wall higher up.

The cliffs (Fig. 3-21), 3–4 m and sometimes 15 m high that occur as risers between 210 and 255 m, are generally smooth with only a few ledges, all of which have rounded outlines (Fig. 3-21B). Although some of the rock walls are very local most can be traced laterally several tens of metres, and in some instances are continuous for distances of at least 50–70 m, the limit of travel. Most cliffs are not vertical, but are inclined at 70–80°. Large shallow water corals such as *Montastraea annularis, Montastraea cavernosa*, and *Diploria* sp. up to 1·5 m across can easily be seen in the rock walls (Fig. 3-21A, C, D). The corals occur in various positions and overturned colonies are not uncommon. On top of many of the cliffs are perched loose heads of *Diploria* 1–2 m across and other coral debris recently arrived from above, identical in size and arrangement to the limestone that forms the cliffs and blocks. The modern rubble looks stained and corroded as though it has not moved very much. The cavities in the walls all have rounded outlines.

It is not clear whether the cliffs are outcrops or giant blocks of rock. Large blocks, some up to 30 m, have acted as dams and sediment has piled up behind them (Fig. 3-22) so that in profile they resemble a sediment tread with a cliff downslope. Only by encircling them is it clear that they are blocks, almost buried by sediment. On the walls of some of these blocks (or possibly outcrops) fronds of *A. palmata* were seen within the limestone.

The sediment that forms most of the slope resembles an alluvial fan or mountain scree because it is so variable both with depth and laterally along the sea floor. Large areas of the bottom are a smooth slope of sand with many gravel-size pieces of *A. cervicornis,* shells and *Agaricia* plates (Fig. 3-23A). Larger coral plates often lie parallel to the face of the slope and if abundant can be shingled. In other areas, either scattered boulders are common, or the entire bottom is a giant field of boulders, mainly rounded heads of coral (Fig. 3-23B), that may have come from the amphitheatre scars above. In some regions there is no sediment on the bedrock and the tops of blocks or outcrops, tens of metres high, often jut up out of the sediment slope and dam large cones of talus behind them (Fig. 3-22). The slope itself is not planar, but when viewed obliquely is formed by a series of broad ridges and swales, with rarely as much as 2 m of relief and many tens of metres wide. Often the coarser material was found in the axes of these shallow swales.

The sediment surface is generally smooth (Fig. 3-23). Any projections that offer

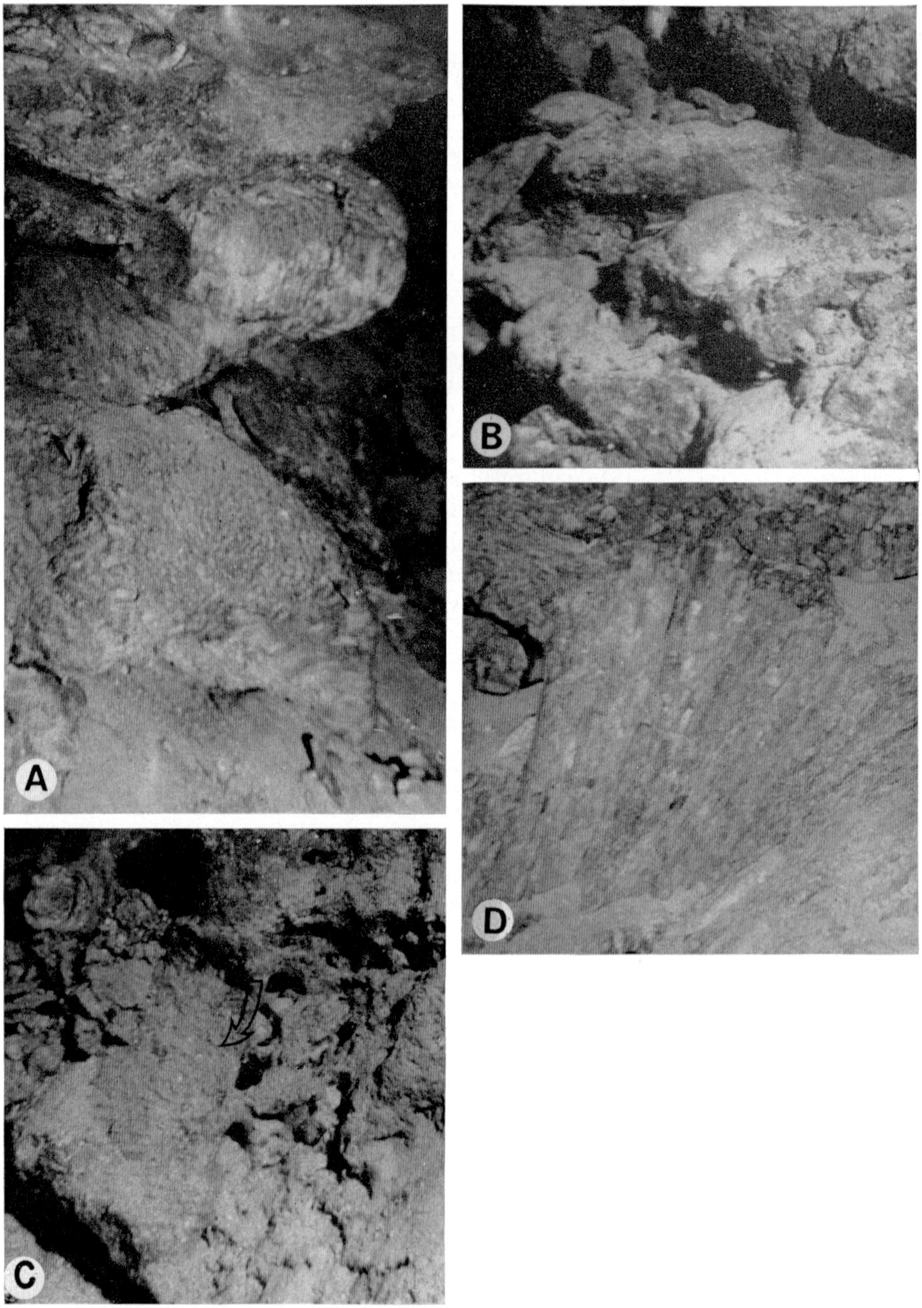

Fig. 3-21. Glovers Reef, site 7, (A) depth 240 m: an oblique view of one of the small cliffs illustrating the makeup of the rock, numerous large shallow-water corals (mainly *Diploria* sp.); note that several colonies are upside down; the colony at the base is about 0·5 m across; (B) depth 205 m: a series of rounded ledges, small caverns and delicate pillars that form the surface of many of the steep walls on the lower reaches of the steep fore-reef off Glovers Reef; the small cave pictured here is about 0·5 m high; (C) depth 215 m: a rock wall made up of what appears to be cemented rubble, partly covered with loose sediment and debris; note the large coral plate at left centre (arrow) cemented

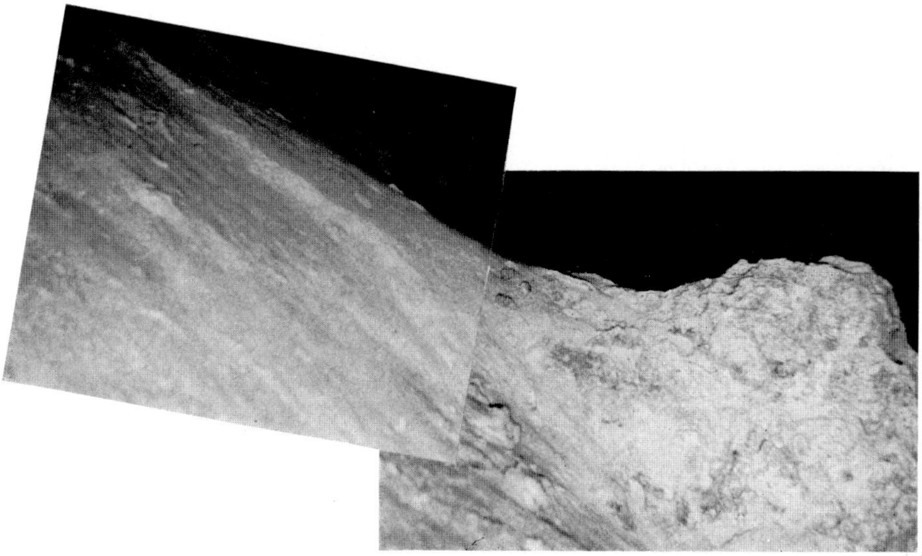

Fig. 3-22. Glovers Reef, site 7, depth 280 m: the smooth slope of fine-grained sediment draped up behind a small block, the block has acted as a sediment dam and the sediment, continuously moving downslope has piled up behind it; the block is about 3 m high.

obstruction to downslope movement have a cone of talus piled up behind them and shadow of fine sand and mud in front. For the most part the sediments are coarse, *Halimeda*-rich sand with many rounded blocks of coral, mostly *M. annularis, Diploria, M. cavernosa,* conchs, and sticks of *A. cervicornis*. Loose *A. palmata* debris was not observed, but pieces of *Millepora* were recognized. There is a surprising abundance of *Strombus* shells that look very fresh. On some areas of the slope, the chutes trending downslope are filled with sediment and rubble; on others they are almost full, indicating episodic movement. It should be noted that blocks, from a few metres to tens of metres across, are common all the way down the slope, from the base of the wall to the deepest depths; they are perched on cliffs and on the smooth slopes almost buried by sediment.

In the lower 30–45 m are several isolated occurrences of coral debris. Particularly interesting are large accumulations of *A. cervicornis* in piles 2–3 m high and 15 by 30 m across. The sticks of coral are cemented together to form hard rock.

In other areas, isolated mounds of *A. palmata* fragments rise up out of a smooth sediment surface slope, these are 3–5 m high, composed of nothing but *A. palmata* (Fig. 3-24D) and may be 7 m wide.

All vertical to near vertical rock surfaces, like those above, are veneered with encrusting organisms. Beginning at a depth of 210 m, one of the most consistent types of encrusters, crustose coralline algae, begins to decrease noticeably; near 225 m, isolated blotches up to 10 cm in diameter rarely cover more than 5% of the available

into the rock; (D) depth 255 m: a close view of the rock surface making up one of the small cliffs on the lower reaches of the cliffed fore-reef, here illustrating a cross-section of a shallow-water coral, *Montastraea* spp. about 30 cm across; to form such a cross-section this part of the wall is clearly the result of fracturing.

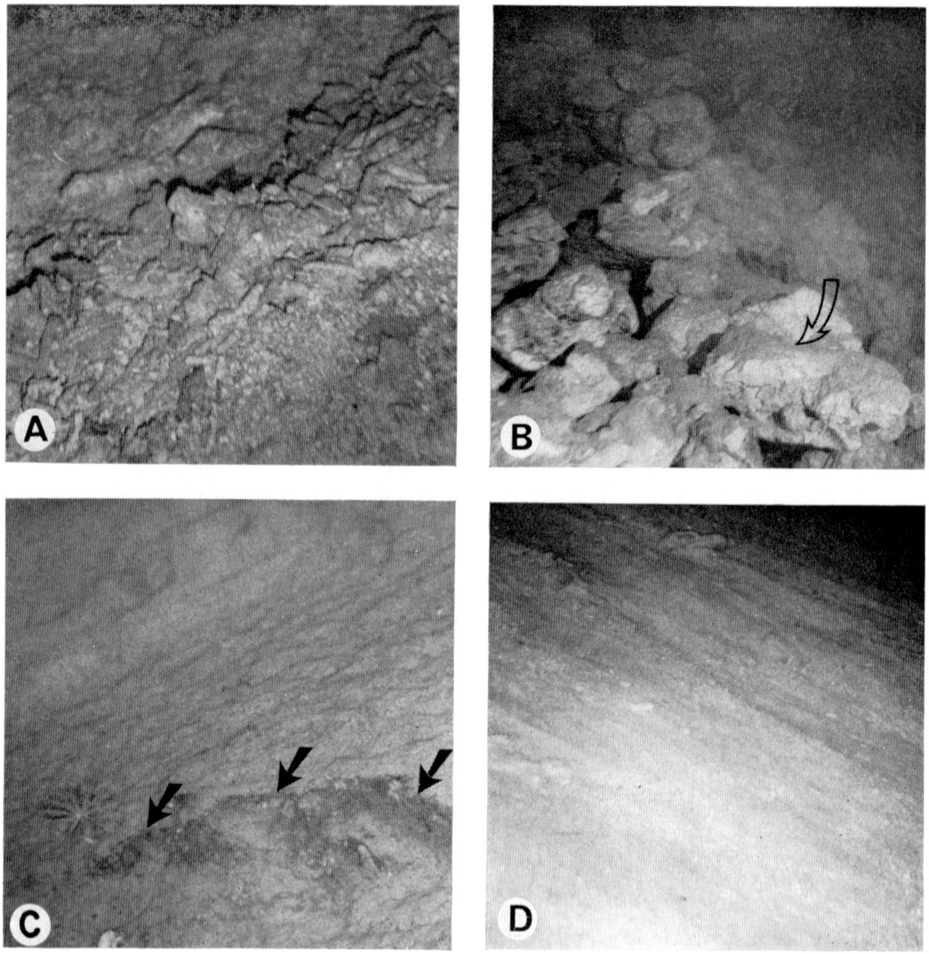

Fig. 3-23. Glovers Reef, site 7, (A) depth 270 m: a slope strewn with coral rubble composed for the most part of *A. cervicornis* sticks and smaller rounded coral heads; the conspicuous sticks are about 5 cm in length; (B) depth 250 m: a close view of a field of coral boulders on a 30° slope; the boulder indicated by the arrow is about 0·5 m across; (C) depth 290 m: the smooth rock surface dusted with fine sediment here illustrating one of the many small mini-ledges (arrows, lower part of photograph) about 0·5 m high and partly overridden with sediment; (D) depth 280 m: the smooth undulating slope of *Halimeda*-rich sand dipping seaward at about 30°.

surface, and the deepest crustose coralline algae were seen at a depth of 250 m. The most common epizoans are still lithistid sponges (Fig. 3-24A) but basket sponges are prolific, and these two forms, along with other unidentified sponges cover up to 50% of the rock surface. Yellow encrusting sponges, so common on the rock walls above, seem to be restricted to the undersides of rocks or indentations. Other common epizoans are eucidarids, cerianthids, yellow and white anemones, crinoids (Fig. 3-24C), blue sponges and tunicates, and especially in deep water (over 300 m) stylasterines (Fig. 3-24B) with ophiroids crawling on the branches.

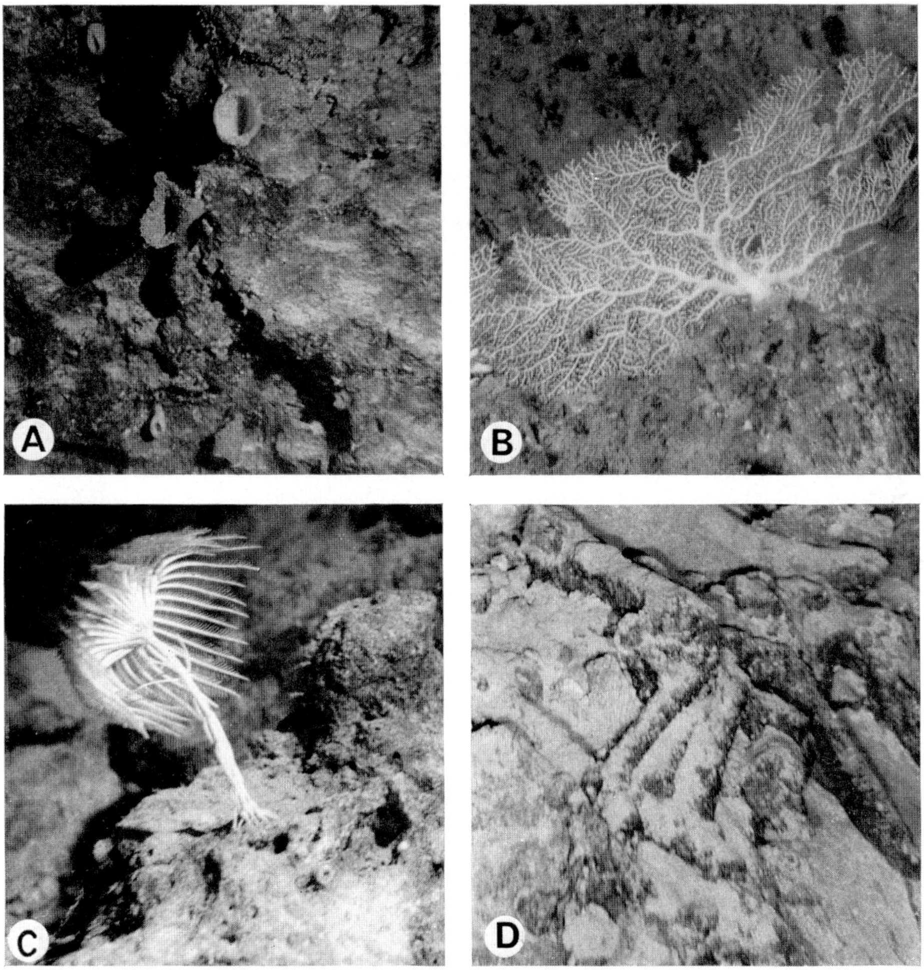

Fig. 3-24. Glovers Reef, site 7, (A) depth 290 m: a smoothly sculptured rock wall with several cup-shaped lithistid sponges, each of which is about 10 cm across growing out from the rock face; (B) depth 240 m: a large stylasterine, about 30 cm across, growing out from a rock wall; (C) depth 270 m: a stalked crinoid *Cenocrinus asterius* about 0·5 m high growing attached to a small rock promontory; (D) depth 300 m: the top of a small rock promontory poking up from the sediment covered surface and composed almost entirely of *Acropora palmata* fronds and sticks, the branches are here about 30 cm in diameter.

Deep precipice (305 m and below)

The relatively gentle 30–40° slope forms the crest of a large escarpment whose top is at about 308–315 m (Fig. 3-25). Below 290 m the smooth surface rock and sediment surface is cut by many large canyons and caves, at widely spaced intervals along the bottom. These canyons have relief of 8–12 m and extend 5–6 m into the rock. At 310–315 m the smooth rock juts out into space, forming the top of a large overhang which is the top of a concave escarpment that extends to unknown depths.

The rock in this lower 15 m is different in appearance from that above, smoother with hardly any megascopic porosity and many straight to angular cracks, or fissures.

Fig. 3-25. Glovers Reef, site 7, depth 310 m: the jagged top of a large escarpment that extends to unknown depths.

SUMMARY

The seaward margin of Belize barrier and atoll reefs is formed by three distinct morphological elements: (1) the reef front, an irregular, seaward-sloping surface covered with hermatypic coral and green algae growth that runs from the reef crest to a depth of between 65 and 70 m; (2) the wall, a vertical to near-vertical cliff from 65–120 m; and (3) the fore-reef, a basinward-sloping accumulation of reef- and wall-derived sediment and talus that grades into a shallow (500 mg deep) trough between the barrier reef and Glovers Reef, or extends as a series of cliffs and steep slopes into the Cayman Trough.

The reef front

The seaward-facing part of the barrier reef is divisible into four separate zones.

The spur and groove (0–20 m). The reef crest extends seaward as a series of coral-covered spurs and sediment-floored grooves from sea level to a depth of about 20 m. In shallow water (less than 10 m), the spurs are built of branched and foliose corals while the grooves are layered with coral rubble and skeletal sand. In deeper water (below 10 m) relief between spurs and grooves decreases noticeably and branched corals predominate on the spurs.

The step (20–35 m). The seaward limit of the spur and groove is abrupt, in the form of a prominent cliff some 15 m high. The upper slopes of this cliff support a luxurious growth of platey corals. The lower slopes have progressively fewer corals and more coralline algae, alcyonarians and sponges.

The sand slope (35–45 m). A plane of coral rubble and skeletal sand corrugated into low-relief swells and swales oriented downslope dips gently seaward from the base of the step. At two of the sites the slope is more reef rock than sediment; (1) an irregular surface covered by scattered living coral and veneered with coral rubble and skeletal sand or (2) a series of large blocks, meters across, upon which a variety of corals and alcyonarians are growing.

The brow (45–65 m). The leading edge of the sand slope gives way gradually to a steeply inclined (45–50°) irregular surface. Mounds or spurs with a luxuriant growth of corals and green algae are separated by depressions or grooves either covered by coral rubble and skeletal sand or floored by reef rock veneered with coralline algae. Coral growth on the upper part of the brow is in the form of large colonies of plate-like corals with many clumps and vine-like growths of deep-water *Halimeda* species. At the transition from the brow above to the wall below, corals cover almost half the available area of the sea floor, a luxuriance that equals that of the shallow water spurs and grooves above.

The wall

The most impressive and recurring feature at all localities is a vertical to overhanging precipice that forms the lower part of the reef zone and extends from a depth of about 65–120 m and occasionally 150 m. On a large scale the surface of the wall consists of ledges of varying size and spacing that are separated by small cliffs or cave-like re-entrants. On a smaller scale the surface is highly irregular with rounded pits, sculptured projections, columns and arches ranging from several centimetres to half a metre in size. The wall is dissected by many near-vertical fissures up to a few metres wide and as much as 30 m high. The bottoms of these fissures are often choked with angular blocks of limestone. In plan view the base of the wall resembles a terrestrial cliff with spurs of rock separated by re-entrants filled with cones of talus.

The near-horizontal surfaces of the wall, ledges, caves and projections, all collect sediment that comes down from the brow and the sand slope above. These surfaces are veneered with a mixture of skeletal sand and mud, in which the large plates of deep-water *Halimeda* sp. are prominent. Whole and fragmented plate-like corals accumulate on the wider ledges, particularly those near the base of the wall.

The upper part of the wall is a biological transition zone, from the shallow-water reef-building community of hermatypic corals, octocorals, green algae, and crustose coralline algae to a deeper water community of coralline algae, octocorals, ahermatypic corals, demosponges, and sclerosponges. Hermatypic corals are common to depths of 70–80 m but rare below these depths; the deepest living specimens were found between 93 and 102 m. Deep-water species of the green algae *Halimeda* are common to 70 m and at one site to 76 m, but they decrease rapidly below these depths; the deepest occurrence of living *Halimeda* was 110 m.

Below the zone of hermatypic corals and green algae, vertical rock surfaces and ledges which are free of sediment support crustose coralline algae and flattish sponges; on the undersides of ledges there are tubular sponges, gorgonians, and hermatypic corals. Sclerosponges are largely cryptic in their habitat growing beneath ledges and in caves; they are common to a depth of 130 m, but were not seen below 150 m.

The fore reef

Two different styles of fore-reef occur along the margins of the Belize complex:

(1) the sloping fore-reef—a relatively gentle slope that decreases in dip rapidly away from the base of the wall and merges with the floor of a shallow trough (400 m) between the barrier reef and Glovers Reef and (2) the cliffed fore-reef—a steep slope composed of cliffs and steeply inclined sediment slopes that forms the western side of the Cayman Trough.

The sloping fore-reef. The accumulation of talus burying the base of the wall and extending seaward resembles a series of coalescing alluvial fans in the mountains. At the base of the wall there is a narrow and steep (35–40°) scree slope composed of blocks of limestone up to 10 m in size in a field of fragments of plate-like corals and coarse-grained sand rich in plates of *Halimeda*. Beyond this narrow zone of talus the slope decreases to about 25° or less, the bottom consists of coarse-grained *Halimeda*-rich sand and the surface exhibits abundant evidence of downslope sediment movement.

The distal portions of the sloping fore-reef are characterized by finer-grained sand, little evidence of active downslope movement and abundant evidence of bioturbation. This transition occurs between 180 and 220 m where the caternary slope flattens to less than 10°. As the slope continues to decrease the fine-grained sand passes into basinal muds rich in pelagic skeletal particles.

At one site, on the leeward side of Glovers Reef, the continuity of the sloping fore-reef is broken by a cliff, some 30 m high, which is either an outcrop or a partially buried, giant limestone block.

The cliffed fore-reef. In this situation the base of the wall is variable from place to place, sometimes occurring at 120 m and sometimes extending to depths of 150 m.

Where the base of the wall is buried by talus at 120 m the blocks of limestone, coral plates and *Halimeda*-rich sand form a very steep (50° or more) sediment slope. The blocks of limestone and coral plates appear to be keyed together and are best termed 'perched talus'.

The bottom between 135 and 210 m is an irregular sequence of small cliffs, 5–20 m high, separated by steeply dipping (45–50°) inclined slopes. The slopes are distinguished by prominent ridges and deeply incised furrows that trend dowslope. The ridges are a cemented talus of massive and platey corals and it is inferred that these rocks have formed like the perched talus above. Between the ridges and against the bases of the numerous cliffs there is a poorly sorted jumble of limestone boulders and heads and plates of coral all surrounded and infilled with sand rich in *Halimeda* plates. Large blocks are sometimes lodged in the furrows and sediment piles up behind them.

The pronounced ridge and furrow topography dies out at about 210 m and the bottom gives way to a series of small cliffs and intervening, sediment veneered rock slopes that dip gently (50–30°) basinward. Fractured surfaces of the cliff walls reveal large colonies of shallow water corals. The rock surface is formed by gentle ridges and swales trending downslope. The sediment veneer is either a conglomerate of loose coral boulders and coral sticks or *Halimeda*-rich sand.

The most common benthic invertebrates in these deeper zones are sponges, either lithistid sponges, basket sponges or other unidentified forms. In addition gorgonians, anemones, crinoids, stylasterenes, eucidarids, cerianthids, ahermatypic corals and tunicates form an important part of the biota. Coralline algae are restricted to surfaces that face the light and decrease in abundance to a depth of 250 m, where the deepest living alga was observed.

Chapter 4

The perireefal sediments

INTRODUCTION

The sediments that surround barrier and atoll reefs, here termed perireefal, are an integral but poorly known element of Holocene reef complexes. The submersible allowed us to observe and photograph the accumulations of sediment from the shallow spur and groove to the distal sloping fore-reef and to collect samples unobtainable by conventional sampling from the sea surface. In addition to collection from the sea floor the numerous samples of wall limestone (see Chapter 5) included samples of unconsolidated and cemented internal sediments. This spot sampling was supplemented by sampling using SCUBA in shallow water and by piston and gravity coring in water too deep for this submersible.

In the previous chapter we have described, in general terms, the sediment associated with various parts of the reef front, wall and fore-reef. This chapter combines our field observations with the results of grain-size analysis and analysis of grain type. Together these results allow us to characterize the different sediment types, determine downslope gradient in grain size and grain composition and lay the groundwork for a later integrated discussion of the source of the sediments and their modes of transport.

SAMPLING

Surface sediments were sampled in two areas along the barrier reef (Fig. 4–1). A complete transect was made near Tobacco Cay and the area to the south was sampled along a shallow-water transect at Gladden Reef and a deep-water transect off South Water Cay.

Where possible, samples were collected directly, using SCUBA equipment to depths of about 45 m, by scooping sediment into plastic bags. Below the depth of SCUBA diving, samples were obtained using a container that, like a Van Dorn water sampler, consisted of a plastic tube 10 cm in diameter and 20 cm long with two sealing caps connected by surgical tubing. The caps were pried off the ends of the tubes prior to diving and tied together with a light cord on the outside of the tube. Four such sampling tubes were tied outside the submersible. On the bottom a tube was grasped with the mechanical claw and a sample scooped from the bottom. Then the light cord holding the closing caps was cut with a razor blade fixed to the back of the mechanical claw allowing the elastic to close the sampler. The filled tube was put in the carrying bag.

ANALYSIS OF SAMPLES

Sediments collected using SCUBA and the submersible were first wet sieved to separate sand and mud size fractions. The sediment coarser than 0·064 mm was then

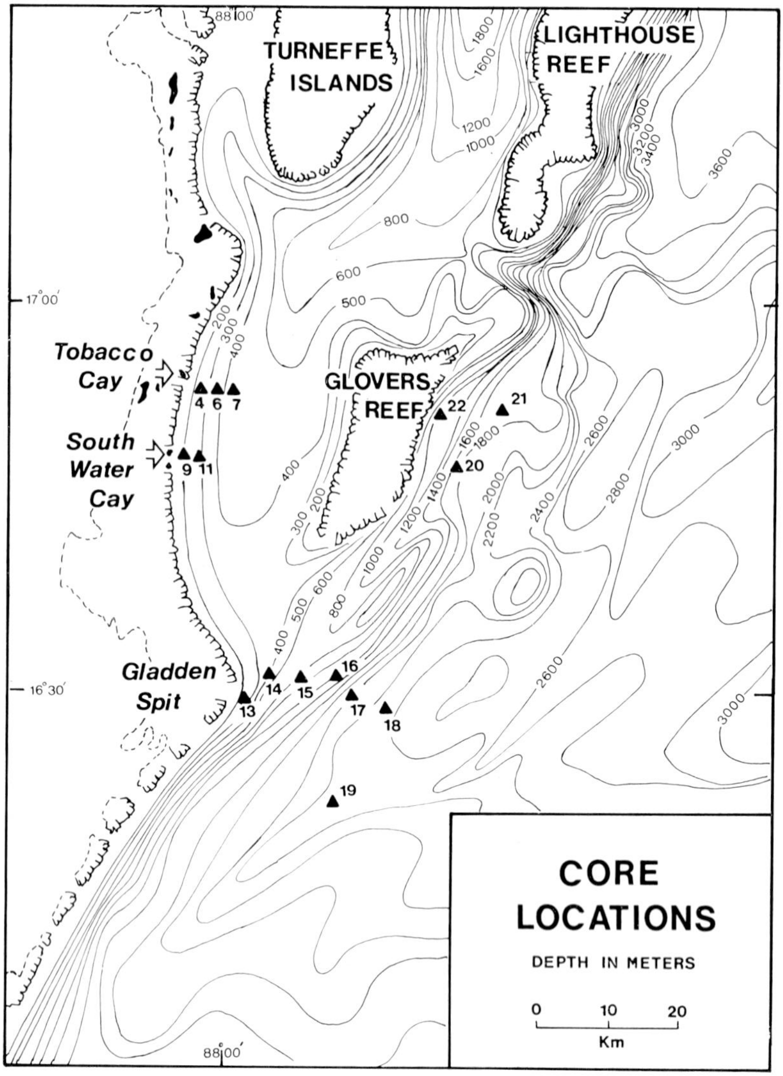

Fig. 4-1. A chart of the southern portion of the Belize barrier reef margin and offshore atolls showing the locations of the piston cores described in this chapter and the location of the two transects of surface sediments at Tobacco Cay and South Water Cay.

air dried and sieved at intervals of 1 φ. The grain-size distribution of the remaining silt and clay-size fractions was estimated with a Coulter Counter by D. S. Marszalek (University of Miami). The composition of each size-fraction in the sand-size range was determined by mounting the grains in plastic, preparing thin sections, staining to differentiate aragonite and Mg-calcite (Winland, 1971; Choquette & Trusell, 1978) and counting 300 grains. The composition of the silt-size fraction was estimated using the scanning electron microscope (SEM).

Cores were logged on board ship by W. J. Koch (Shell Oil Co.), sealed, kept moist and then sampled later in the laboratory. Samples were taken selectively to represent the different sediment types down the core, and were analysed for grain size as above. The composition of the coarser-grained sediments was examined in thin-section by W. J. Koch and M. Brady (University of Kansas). Variations in the mineralogy of the cores were also determined by Brady (1974) using X-ray diffraction and thin section petrography.

SEDIMENTS OF THE SHALLOW REEF

The shallow crest, whether along the barrier reef or on Glovers Reef, forms a ridge of living coral that separates the reef front from the reef flat. The reef flat, from 1·5 to 10 km wide and rarely more than 4 m deep, is mostly covered with white sand swept into sand waves oblique to the trend of the reef; locally these sands are stabilized by the sea grass *Thalassia testudinum*. Directly in the lee of the crest is the pavement, a zone 10–100 m wide of cemented reef rubble with some scattered coral growth, dusted with skeletal sand and strewn with coral boulders transported across the reef during storms (James *et al.*, 1976).

Surface sediments

Most of the loose surface sediment of the shallow reef margin, in the grooves, on the pavement and on the sand apron, is coarse-grained. Sediments adjacent to the reef, on the pavement or in shallow grooves, are coarsest with a conspicuous granule fraction (Fig. 4-2). This granule fraction, reflecting nearness to the reef, is composed

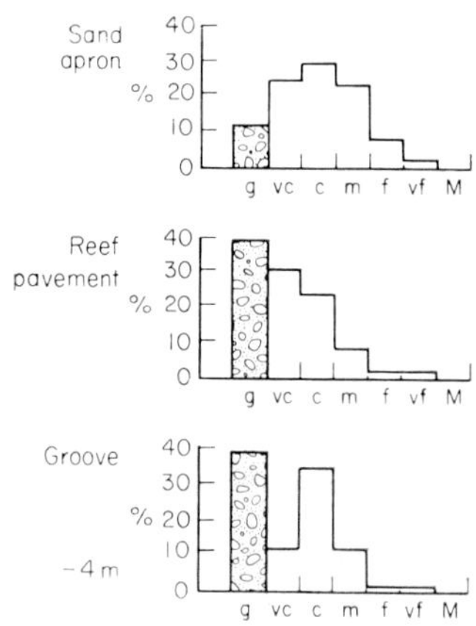

Fig. 4-2. The grain-size distribution of surface sediments by weight from samples taken in shallow water near the reef crest at Tobacco Cay: G, granule size (greater than 2 mm); vc, very coarse sand size (2–1 mm); c, coarse sand size (1–0·5 mm); m, medium sand size (0·5–0·25 mm); f, fine sand size (0·25–0·125 mm); vf, very fine sand size (0·125–0·064 mm); M, mud (less than 0·064 mm). Each plot is the mean of two different samples.

of coral fragments (and branches), gastropod shells, *Millepora* fragments and whole *Homotrema* tests (Fig. 4-3A, B). The sediments on the sand apron, in the lee of the pavement, are predominantly coarse to very coarse-grained sands. Everywhere in this area the sand-sized fraction comprises *Halimeda* plates, coral fragments, pieces of coralline algae, broken tests of the encrusting foraminifers *Homotrema* and *Gypsina* and echinoid spines and plates.

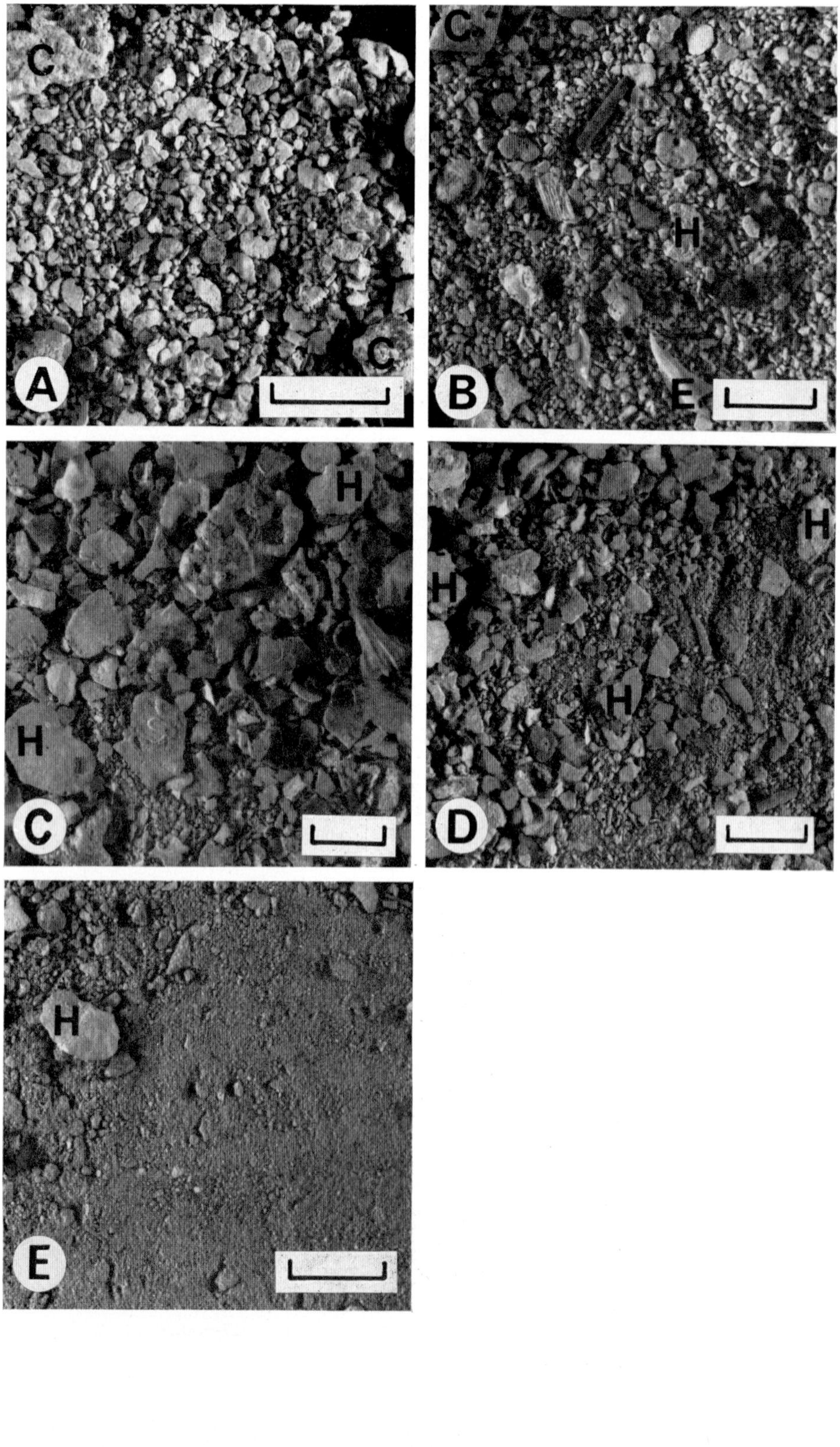
C
C
A
C
H
B
E
H
H
C
H
H
H
D
H
E

Internal sediments of the spurs

In shallow water, whenever coral-covered spurs were pried open searching for lithified sediment the water quickly turned murky with the fine-grained sediment pouring out of the structure (James *et al.*, 1976). Although none of this sediment was sampled, similar internal sediments that have been lithified both in spurs and in the pavement were studied in detail. This cemented sediment has a packstone (Dunham, 1962) texture and is composed of gravel and sand-sized skeletal grains surrounded by a matrix of silt that is cemented by Mg-calcite. The sand-size fraction consists of about half *Halimeda* plates and coral fragments, and half fragmented tests of the encrusting foraminifers *Homotrema* and *Gypsina*, other benthic foraminifers, echinoid spines and coralline algae. The silt comprises about equal proportions of peloids of Mg-calcite micrite and fragments of aragonite (coral) produced by endolithic sponges.

SEDIMENT OF THE LOWER REEF FRONT AND WALL

The sediments of the wall were sampled in four places: (1) on a ledge projecting out from the wall at 92 m (South Water Cay) (Fig. 4-3C); (2) the floor of a small cave some 2 m in from the surface of the wall at 100 m (Tobacco Cay); (3) an excavation into the brow at 48 m (Glovers Reef), and (4) some 4 m in from the surface at the same locality. At Tobacco Cay and South Water Cay, samples were collected by submersible; off Glovers Reef, divers collected both coral and surrounding loose sediment. As soon as the artificial exposure was touched during sampling, clouds of mud poured out of the exposure and obscured vision within a few moments.

The grain size of all four samples is plotted on Fig. 4-4 and the relative proportion of constituent particles in each size fraction of these two samples is plotted on Fig. 4-5. The total composition of these samples is tabulated in Table 4-1.

The sediment from the open ledge is the coarsest, with more than half of the particles granule size and larger (Fig. 4-4), mostly *Halimeda* plates and some coral fragments. Those sediments from the protected, interior part of the reef front and the wall have fewer granule-size particles, about the same proportion of sand-sized particles and much more mud than sediments on the ledge (Fig. 4-4).

All sediments have three distinct size modes; granule, coarse to fine sand, and mud (Fig. 4-4). Judging from the two samples analysed (Fig. 4-5) the granule-size particles are mostly clasts and *Halimeda* plates. The percentage of clasts is misleading, however,

Fig. 4-3. (A) Sediment sample from the reef pavement at Tobacco Cay, just on the lee of the reef crest in water less than one metre deep; composed of rounded coral clasts [C], *Halimeda* plates [H] and skeletal sand: scale bar 1 cm. (B) Sediment sample from a groove between two coral covered spurs, running seaward at a depth of 5 m off Tobacco Cay composed of coarse particles of coral [C], *Halimeda* [HA], *Homotrema* [H] and echinoid spicules [E] in medium to coarse grained sand; scale bar 1 cm. (C) Sediment sample taken from a small ledge on the wall at a depth of 92 m, off South Water Cay, site 2 (sample No. 28.4) and composed of many large plates of the calcareous green alga *Halimeda cryptica* [H]; scale bar 2 cm. (D) Sediment sample taken from near the base of the wall at the top of the sloping fore-reef off Tobacco Cay, site 1, at a depth of 116 m (sample No. 30.2) comprising many large granule size plates of *Halimeda* [H] and fine to medium-grained skeletal sand; scale bar 2 cm. (E) Sediment sample taken at the transition between proximal and distal portions of the fore-reef at a depth of 190 m off Tobacco Cay, site 1 (sample No. 30.4) composed of only a few *Halimeda* plates [H] and much fine-grained skeletal sand; scale bar 2 cm.

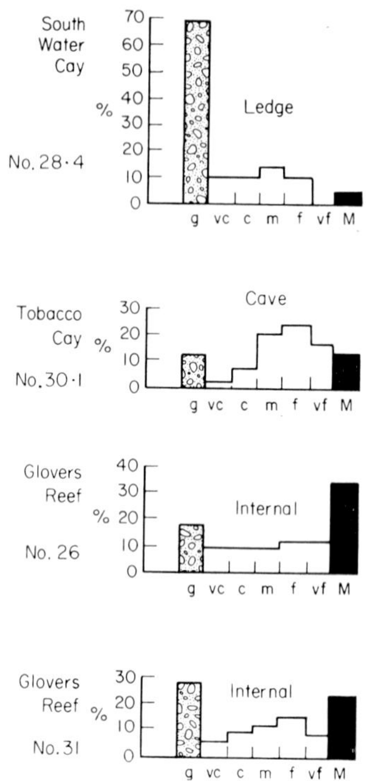

Fig. 4-4. The grain-size distribution, by weight, of sediments from inside a rock sample site (Nos 26 and 31; depth 48 m) on the brow at Glovers Reef, from a ledge on the wall at a depth of 92 m (No. 28.4) and from inside a small cave in the wall at a depth of 100 m (No. 30.1); for locations see Fig. 4-1.

because in the artificial exposure these are poorly lithified sediment broken off during sampling with explosives and not true sedimentary clasts. The sand-size mode is made up mainly of fragmented *Halimeda* plates, mollusc shells, coral fragments, coralline algae grains, peloids, whole and fragmented benthic foraminifer tests, and in the fine and very fine sand grades, the chips eroded by endolithic sponges. The composition of the mud is discussed separately later in this chapter.

There are significant differences in the relative proportions of skeletal constituents in sediment from beneath the reef rock surface at the brow (40 m) and in sediment in a cave on the wall (100 m). The loose sediment between coral heads 4 m beneath the growing reef surface contains a higher proportion of both coral fragments and chips eroded by boring sponges. In contrast, the sediments on the floor of a cave near the base of the wall contain almost twice as much *Halimeda* (reflected as a larger granule size fraction) and almost three times as many benthic foraminifers. The proportion of mollusc fragments, coralline algae grains, echinoid pieces and peloids is about the same in both samples and together these grains comprise roughly a fifth of the total sediment. The differences likely reflect the location and nature of the sedimentary environment; the high proportion of coral and sponge chips reflects the overlying, living reef in shallow water while the abundant *Halimeda* and benthic foraminifers probably reflect the composition of the sediment that moves down the wall.

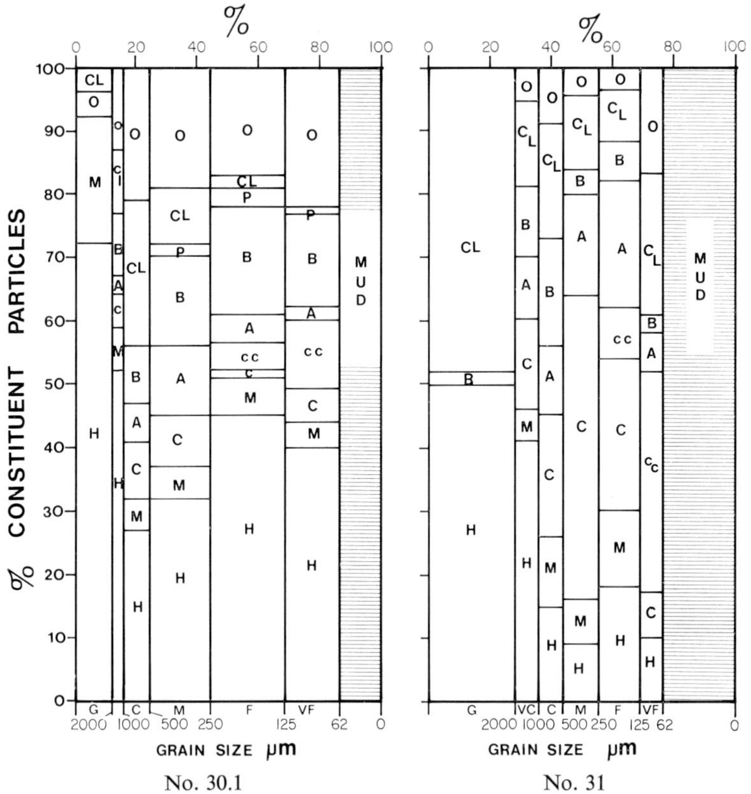

Fig. 4-5. Diagrams (Neumann, 1965) depicting the percentage of constituent particles in each size fraction. The width of the columns represents the relative weight percentage of any particular size fraction of sediments (right side) inside a sample site on the brow (No. 31; depth 40 m) off Glovers Reef and (left side) inside a small cave on the wall (No. 30.1, depth 100 m) off Tobacco Cay. The percentage composition was determined by counting; H, *Halimeda*; M, mollusc; C, coral; cc, sponge chips; A, coralline algae; B, benthic foraminifers; P, planktonic foraminifers; CL, intraclasts; O, other.

SURFACE SEDIMENTS OF THE SLOPING FORE-REEF AND BASIN

As seen from the submersible, there is a rapid and dramatic decrease in grain size of the surface sediments from the proximal part of the fore-reef (scree of boulders, blocks and sand) to the distal part of the fore-reef (fine-grained sand); this trend of decreasing grain size continues into the basin where mud-size sediment prevails. These decreases in grain size are shown in Fig. 4-6 where the samples are grouped by zone. Sediments on the proximal fore-reef are generally bimodal with one mode in the granule size and a second mode in the coarse to fine sand size, and less than 5% mud. Sediments in the transition zone from proximal to distal fore-reef have smaller percentages of granules (maximum 12%), more fine-grained sand and more mud (20–30%). The two samples from the distal fore-reef have only the finest sand sizes and 60% mud; sediments from the basin are over 70% mud.

Parallel to the downslope decrease in grain size there are changes in the constituent composition. The major changes in the total composition downslope on the

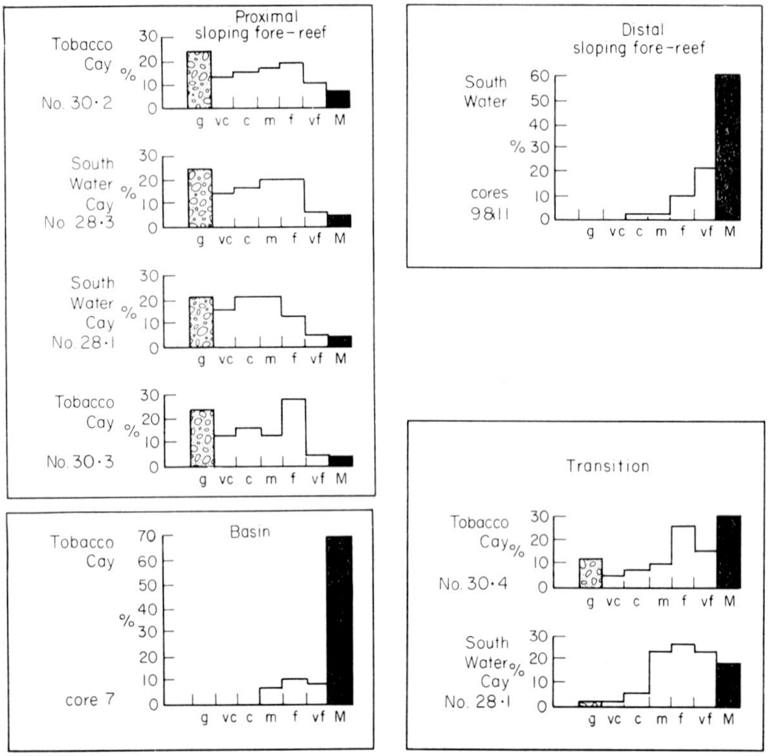

Fig. 4-6. The grain-size distribution of surface sediments, by weight, from the proximal and distal fore-reef as well as basinal environments. Depth of samples: No. 30.2 = 116 m; No. 28.3 = 120 m; No. 28.1 = 168 m; No. 30.3 = 177 m; No. 30.4 = 190 m; No. 28.1 = 200 m; core 9 = 216 m; core 11 = 253 m; core 7 = 400 m.

proximal part of the fore-reef are a decrease of about 20% in the proportion of *Halimeda* and a four-fold increase in the relative percentage of sponge chips (Table 4-1 and Fig. 4-7). This distribution suggests downslope sorting during sediment movement; *Halimeda* is left high on the proximal fore-reef, coarse to fine-grained sand with many different components is transported downslope, and very fine-grained sponge chips are carried in suspension to the base of the proximal fore-reef.

Although no samples from the distal fore-reef or basin environments were counted, visual estimates of the composition of the sand-size fraction indicate that this trend of changing composition continues. In the distal fore-reef the sediments are *Globigerina*-rich sands and silts. The sand-size fraction is estimated to contain one quarter pelagic elements (foraminifers 20%, pteropods 5%), one quarter skeletal debris from shallow-water organisms and one quarter benthic foraminifers (miliolids, peneropolids, rotalids). The remaining grains are either chips from endolithic sponges (15%) or ovoid faecal pellets (10%).

The variations in the relative percentages of constituents with grain size are shown in Figs 4-7 and 4-8. The percentages of most of the constituents do not change markedly in the various size fractions; *Halimeda,* coral, mollusc, coralline algae and to a somewhat lesser degree benthic foraminifera particles are all conservative. The percentage of sponge chips is inversely proportional to grain size, and the percentage

Table 4-1. Grain composition of sediment samples from the reef wall and sloping fore-reef

Sample No. / Depth	Samples: Location	Percentage of various particles*															
		Halimeda	Mollusc	Coral	Coralline algae	Echinoid	Benthic foram	Pelagic foram	Sponge chips	Worm tubes	Intraclasts	Peloids	Sponge spicules	Ascidian spicules	Alcyonarian spicules	Clay-size carbonate (< 4 μm)	Unknowns
72-31 48 m	Soft sediment- 4 m inside reef-wall Glovers Atoll—Site 7	23	4	12	7	1	4	2	11	—	20	6	< 1	1	—	2	7
72-30.1 100 m	Reef wall—inside cave Tobacco Cay—Site 1	39	6	6	6	3	11	4	7	2	6	5	< 1	tr	tr	1	4
72-28.3 120 m	Top of sloping fore-reef South Water Cay—Site 2	50	6	11	6	3	9	4	1	2	4	2	< 1	tr	< 1	< 1	1
72-30.3 177 m	Sloping fore-reef Tobacco Cay—Site 1	53	4	6	6	2	13	2	3	—	3	2	tr	tr	tr	1	5
72-30.4 190 m	Sloping fore-reef Tobacco Cay—Site 1	29	5	8	6	2	11	4	12	< 1	2	9	tr	tr	tr	3	9
72-28.1 200 m	Sloping fore-reef South Water Cay	35	3	7	3	tr	10	7	12	2	1	6	< 1	< 1	< 1	2	11

*The percentage of various particles in the granule, sand and silt size fractions of the sediment samples was calculated by multiplying the relative percentage of any given particle type in each size fraction by the weight percent of the sample occupied by the grain size fraction and summing the result.

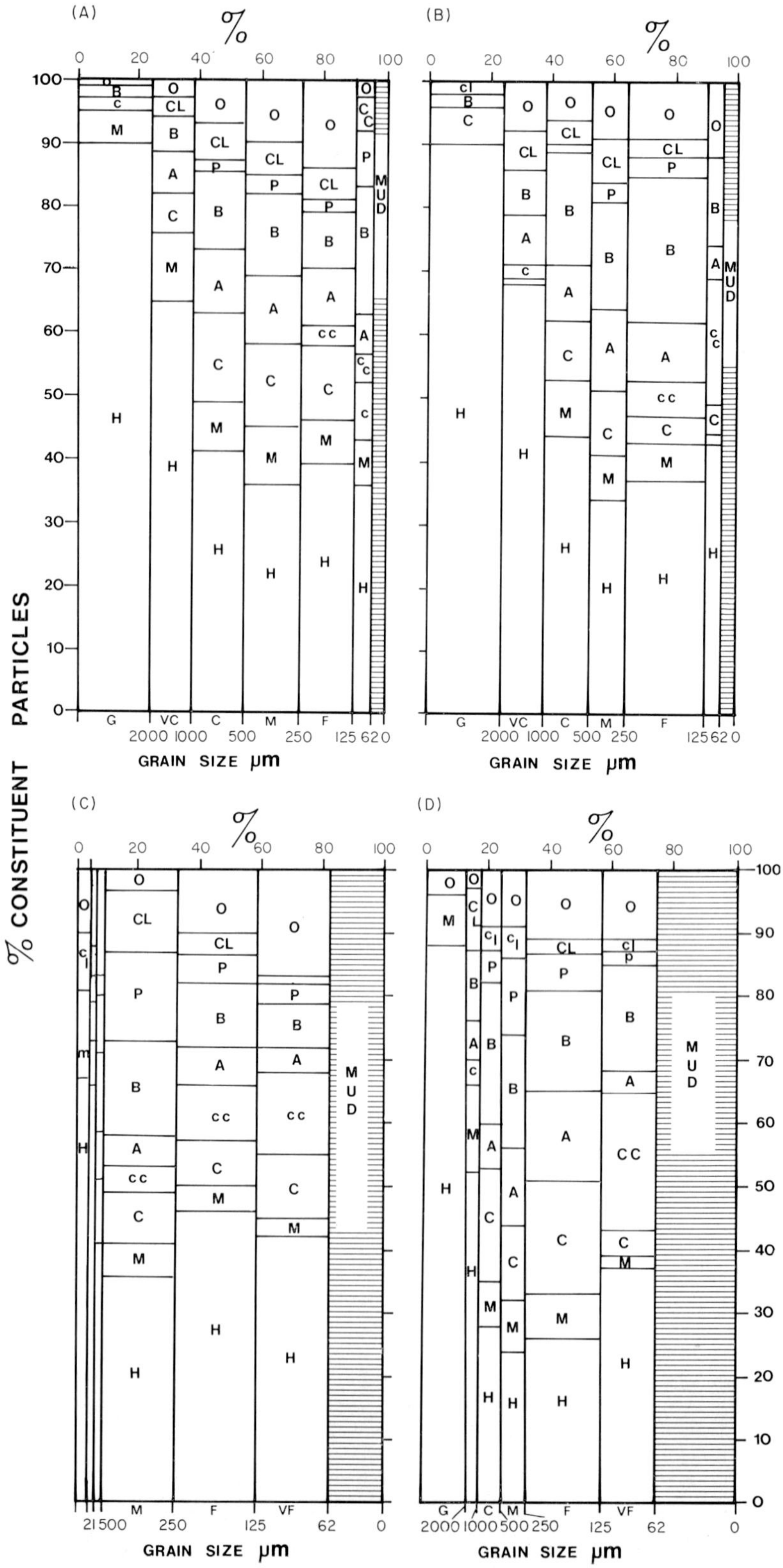
(A)
(B)
(C)
(D)
%
% CONSTITUENT PARTICLES
GRAIN SIZE μm
MUD

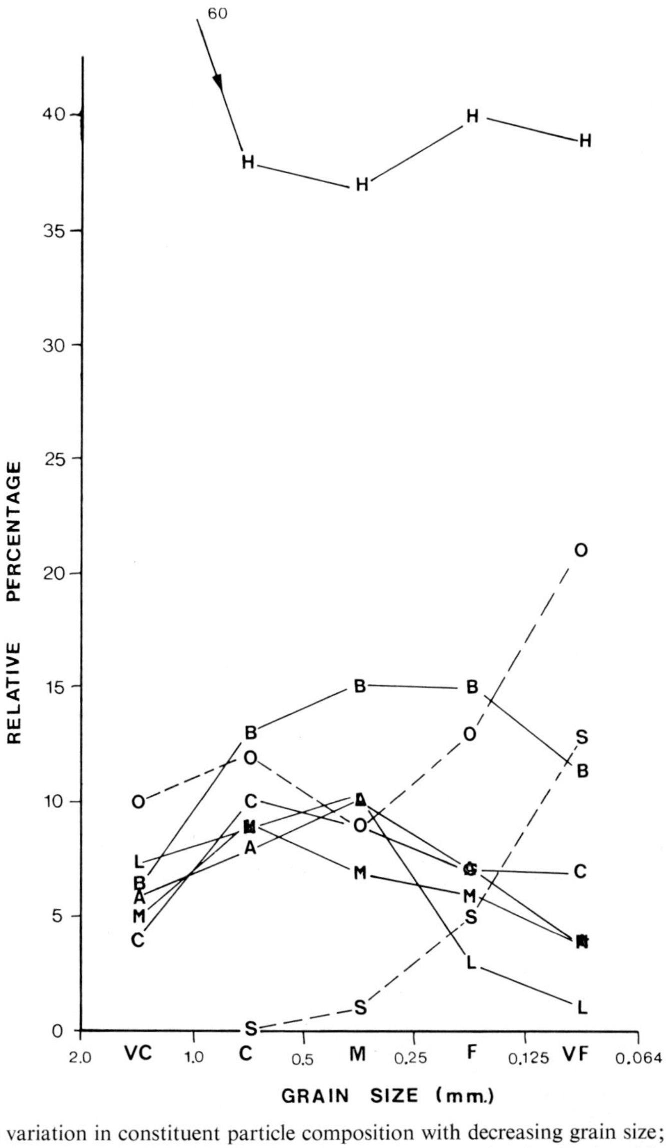

Fig. 4-8. The variation in constituent particle composition with decreasing grain size; an average of six sediment samples (Nos 28.1, 28.3, 30.1, 30.4, 31). H, *Halimeda*; B, benthic foraminifera; L, lithoclasts; A, coralline algae; M, mollusc; C, coral; S, chips eroded by endolithic sponges; O, other (unidentified).

Fig. 4-7. Diagrams (Neumann, 1965) depicting the percentage of constituent particles in each size fraction, with the width of the columns representing the relative weight percentage of any particular size fraction of sediments from the (A) proximal, No. 28.3; depth 120 m (B) proximal, No. 30.3; depth 177 m, and (C) transitional, No. 28.1; depth 200 m, (D) transitional, No. 30.4; depth 190 m, parts of the sloping fore-reef. The percentage composition was determined by counting; H, *Halimeda*; M, mollusc; C, coral; cc, sponge chips; A, coralline algae; B, benthic foraminifers; P, planktonic foraminifers; CL, intraclasts; O, other.

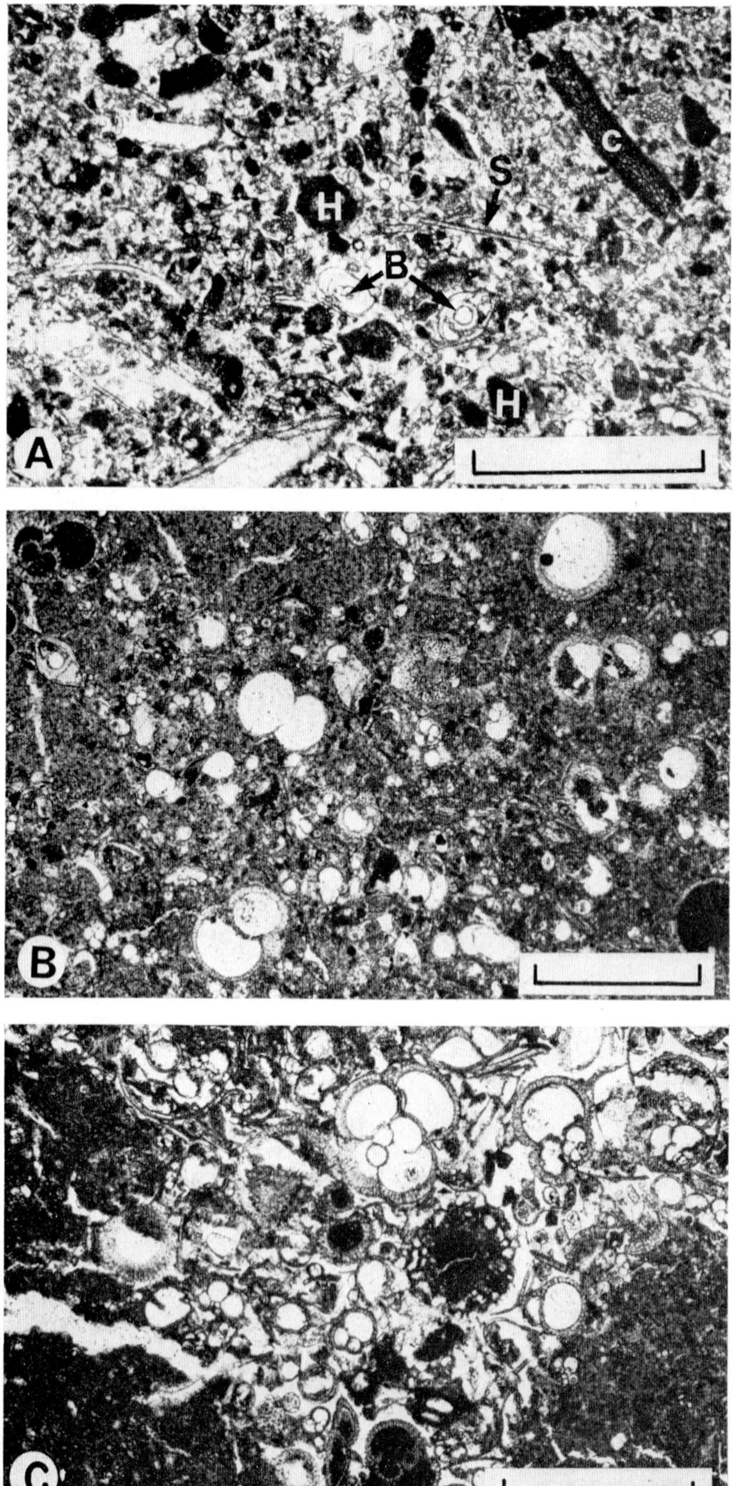
C
S
H
B
H
A
B
C

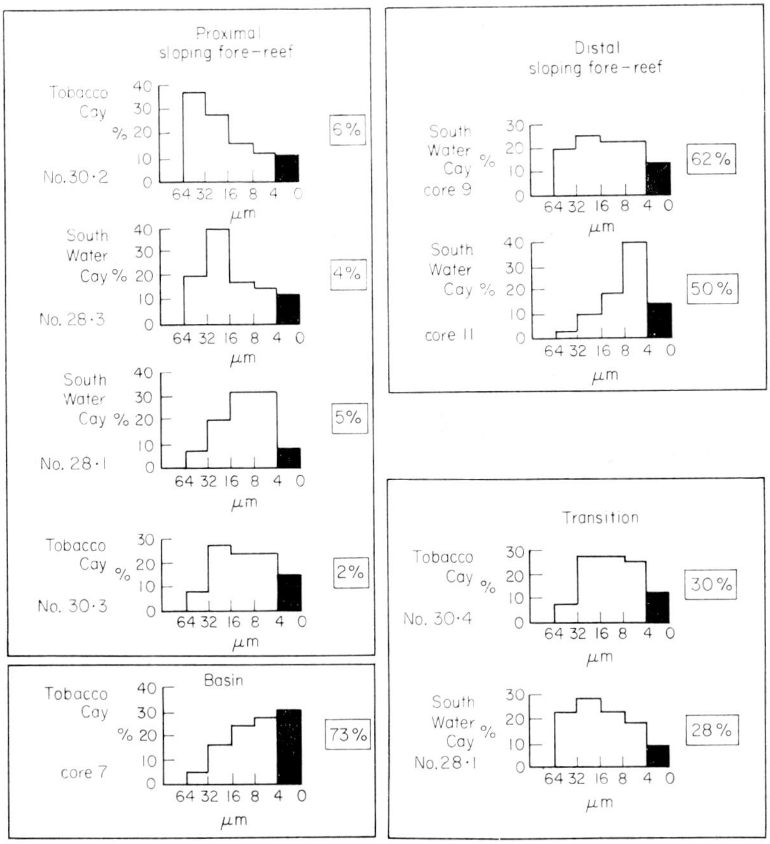

Fig. 4-10. The grain-size distribution of mud in surface sediments from the proximal, transitional and distal parts of the sloping fore-reef as well as the basin, as determined by Coulter Counter; the number in the box beside each plot is the total percentage of mud in that particular sample. Depth of samples; No. 30.2 = 116 m; No. 28.3 = 120 m; No. 28.1 = 169 m; No. 30.3 = 177 m; No. 30.4 = 190 m; No. 28.1 = 200 m; core 9 = 218 m; core 11 = 253 m; core 7 = 400 m.

of clasts is directly proportional to grain size. These differences are hereditary, sponge chips are mainly less than 0·1 mm and clasts are characteristically coarse sand sized and larger.

The basinal sediments consist predominantly of pelagic elements with only occasional benthic foraminifers in mud with no recognizable grains from shallow-water reef-associated organisms.

Most grains are fresh and show little alteration. Synsedimentary cement is common

---

Fig. 4-9. (A) Core 9, site 3 (South Water Cay) depth 220 m: a photomicrograph in plane-polarized light of *Globigerina* silt from the distal part of the sloping fore-reef, here illustrating a sample rich in fine-grained, shallow-water derived material particularly coralline algae [C], sponge spicules [S], benthic foraminifers [B] and *Halimeda* fragments [H], in a matrix of mud; scale bar 500 μm. (B) Core 7, site 1 (Tobacco Cay), depth 400 m : a photomicrograph in plane polarized light of *Globigerina* mud from the basinal zone comprising planktonic foraminifers with empty tests (open circles) in mud; the dark circles are planktonic foraminifer tests filled with Mg-calcite micrite cement; scale bar 500 μm. (C) Core 7, site 1 (Tobacco Cay), depth 400 m: a photomicrograph in plane-polarized light of the same *Globigerina* mud as in (B) above that has been burrowed and the burrows filled with planktonic foraminifer tests; scale bar 500 μm.

in the utricles of *Halimeda* plates, in gastropod chambers and in the pore spaces between silt-sized carbonate grains of agglutinated foraminifer tests. Aragonite cement is common in the utricles of *Halimeda* and the chambers of gastropods, Mg-calcite is commonest in foraminifer tests but is also often encountered in *Halimeda* utricles. Even though many other skeletal grains have internal pores they rarely exhibit any internal cement. The periphery of most grains is bored by endolithic algae but the holes are rarely filled. Occasionally *Halimeda* plates are encrusted with coralline algae or the tests of the encrusting foraminifer *Gypsina*.

Halimeda is the major constituent of the proximal and transitional zones; the species present are mostly *H. cryptica, H. goreauii* and lesser amounts of *H. copiosa,* all of which grow only on the deeper parts of the reef (brow) and at the top of the wall. Shallow-water species of *Halimeda* are rare. The consistency in the percentage of *Halimeda* in the various sand-size fractions is the result of the ease with which segments are fragmented. Each segment has an axial void that allows it to separate into two halves of approximately the same size as the entire segment. Pores (utricles) that are perpendicular to the segment surface facilitate breakdown to fine sand sizes.

MUD

The percentage of mud in the surface sediments increases rapidly with increasing depth: proximal sloping fore-reef, less than 6%; transitional zone between proximal and distal portions of the sloping fore-reef average about 30%; distal sloping fore-reef, 50%–70%; basin, 73% (Fig. 4-10). Remarkably, the relative percentage of the clay-size fraction is less than 15% of the total mud fraction, regardless of the total amount of mud; only in the sample from the basin does the relative percentage of clay-size carbonate approach 30% (Fig. 4-10). Most of the silt-size fraction falls between 4 and 32 μm and except for the sample from shallowest depth (Tobacco Cay transect, 116 m, Fig. 4-10) the shapes of the histograms suggest that the silt is not closely related to the sand-size fraction. The mud fraction of internal sediments (Fig. 4-11), shows the same general pattern as that of the surface sediment with one noteworthy exception: the sample from the ledge off South Water Cay, while containing much less mud, has almost 30% clay-size carbonate in the mud fraction making it resemble the basin sediment.

Qualitative examination of the silt-size fraction, in thin section (coarse silt) and under the scanning electron microscope (medium to fine-grained silt) of all the samples in Fig. 4-10, indicates that the silt is composed of a wide variety of grains that fall into four general categories: (1) obvious skeletal grains (20–40%) of which fragments of echinoids, *Homotrema,* planktonic foraminifers, benthic foraminifers, sponge spicules and ascidian spicules can be recognized, (2) angular grains that are brown in thin section and resemble *Halimeda* (10–30%), (3) unidentifiable grains and clasts, usually composed of Mg-calcite (20–40%), (4) chips of aragonite eroded by endolithic sponges (30–40%).

When combined with the relative percentage of sponge chips observed in the very fine sand fraction of the same samples, these grains are calculated to form 7–11% of the total internal sediment in the reef front and wall, less than 3% of the sediment on proximal parts of the fore-reef, and as much as 12% of the total sediment in fine-grained sediments from distal fore-reef and basinal settings.

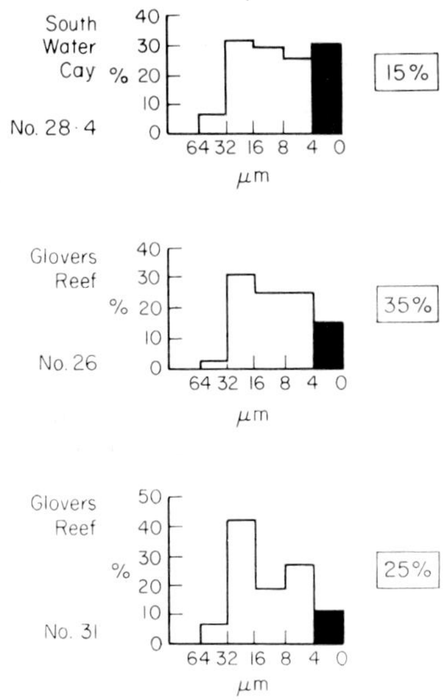

Fig. 4-11. The grain-size distribution of mud sediments from a rock sample site on the reef brow of Glovers Reef (No. 26, depth 48 m and No. 31, depth 48 m) and from a ledge on the wall off South Water Cay (No. 28.4, depth 92 m) as determined by Coulter Counter; the number in the box beside each plot is the total percentage of mud in that particular sample.

SUBSURFACE SEDIMENTS ON THE DISTAL FORE-REEF AND BASIN

We tried to take cores of the proximal fore-reef but none of the piston cores penetrated more than a few centimetres. This resistance is probably the result of the coarse sands in depths shallower than 200 m, or possibly attributable to cemented sand below the surface. We were, however, able to take cores of the distal fore-reef and basin environments.

Five cores were taken between 210 and 400 m out into the basinal trough: three cores off Tobacco Cay and two cores off South Water Cay (Fig. 2-1); the deepest core at 400 m is 4·1 km seaward of the wall. The lithology and size frequency histograms for major sediment types in these cores are plotted graphically in Figs 4-14 and 4-15. Most cores penetrated less than one metre below the sediment surface but one core off South Water Cay penetrated 4·9 m subsurface, yielding a good cross-section of the sediment close to the transition from fore-reef to basin.

The sediments found in the cores are of three different types: (1) *Globigerina*-rich sands and silts, green-grey in colour and similar to those sediments on the surface in the distal fore-reef (Fig. 4-9), (2) terrigenous silt, mainly quartz and feldspar, and (3) intraclasts (Fig. 4-13A), brown to grey in colour and composed mainly of pelagic components.

The intraclasts are irregular in shape and range in size from 0·5 to 3·0 cm. They are composed of two lithologies: (1) *Globigerina* mudstones, similar to the sediments of the distal fore-reef and basin but well cemented with Mg-calcite micrite (Fig. 4-13B),

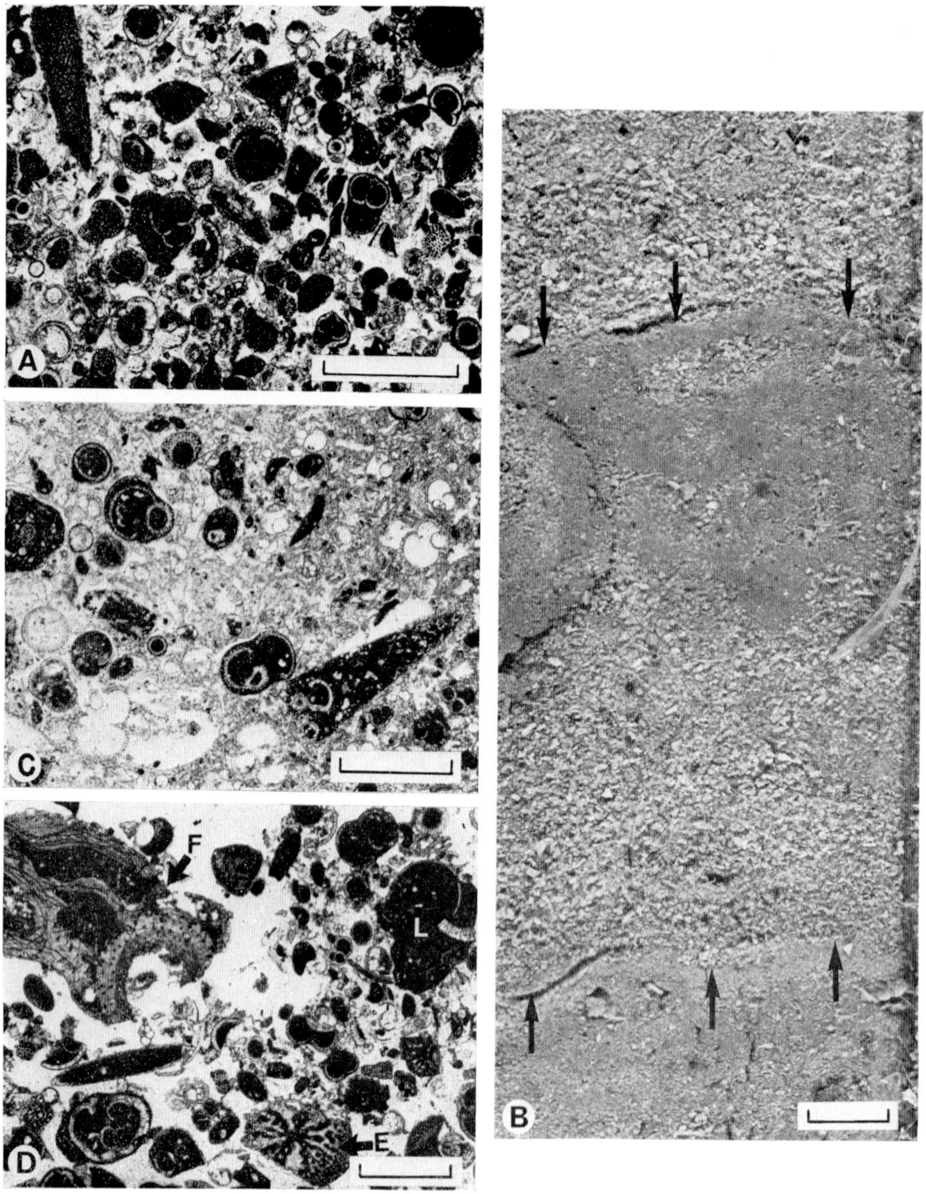

Fig. 4-12. (A) Core 11, site 2 (South Water Cay) depth 350 m: photomicrograph under plane-polarized light of muddy *Globigerina* sand in which most of the foraminifer tests are filled with Mg-calcite cement; scale bar 500 μm. (B) Core 6, site 1 (Tobacco Cay), depth 350 m: a split section of core illustrating one of the sand-layers (between arrows) that grades from skeletal sand at the base to muddy *Globigerina* sand at the top; scale bar 1 cm. (C) A photomicrograph under plane-polarized light, of the muddy *Globigerina* sand at the top of the graded layer illustrated in **(B)** above, some of the globigerinid foraminifers and pteropods are filled with Mg-calcite micrite, others are empty; scale bar 500 μm. (D) A photomicrograph under plane-polarized light of the skeletal sand at the base of the graded layer composed of mixed skeletal elements from the distal fore-reef comprising encrusting foraminifers [F] echinoid spines [E] and lithoclasts [L] along with planktonic foraminifer tests filled with Mg-calcite micrite cement, some of which are fragments; scale bar 500 μm.

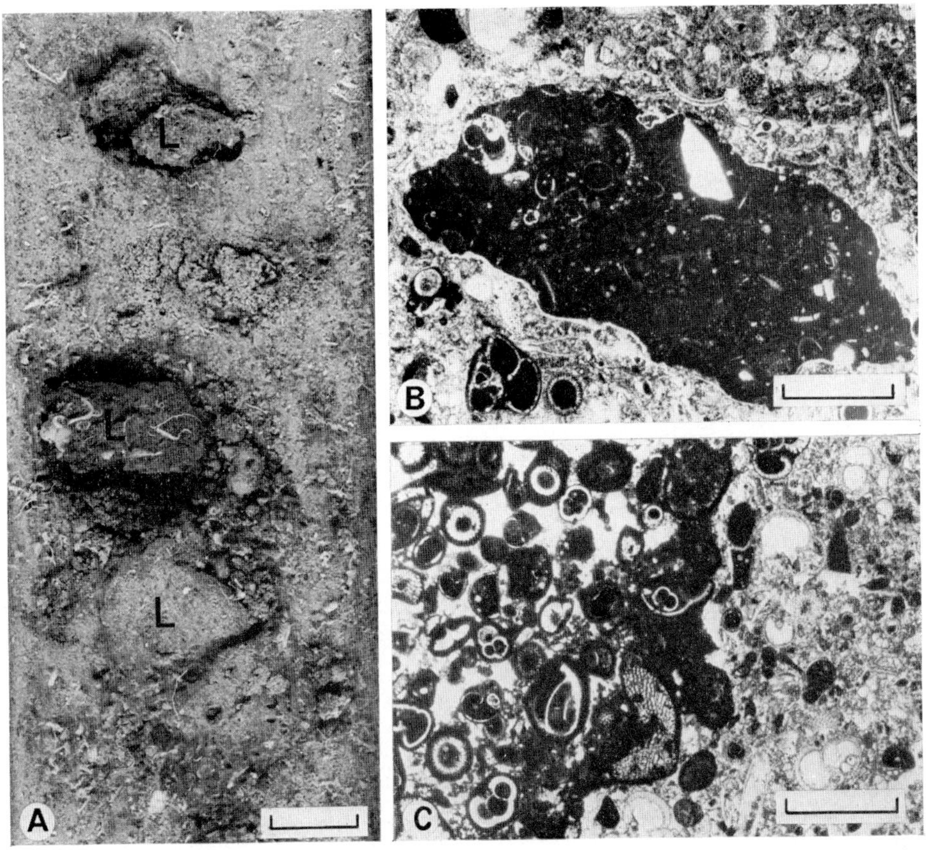

Fig. 4-13. (A) Core 6, site 2 (South Water Cay), depth 350 m; a split section of core illustrating intraclasts (L) surrounded by soft sediment, note the white serpulid worm tubes on the dark clast, centre; scale bar 1 cm. (B) A photomicrograph under plane-polarized light, of a small lithoclast composed of planktonic foraminifer mud, cemented by Mg-calcite micrite size cement; scale bar 500 µm. (C) A photomicrograph under plane-polarized light of the margin of another lithoclast (left) composed of planktonic foraminifer lime grainstone in which each particle is infilled with Mg-calcite micrite cement, and cemented to other particles by Mg-calcite micrite; scale bar 500 µm.

(2) *Globigerina*-rich grainstone. The grainstones are composed of medium to fine-grained skeletal particles, half of which are pelagic foraminifers and half of which are echinoid, coral, mollusc, benthic foraminifer and pteropod grains (Fig. 4-13). The chambers of the foraminifers and voids in other skeletal particles are filled with Mg-calcite micrite cement. The rock is friable, with particles poorly cemented by Mg-calcite micrite (Fig. 4-13C). These grainstone intraclasts apparently disaggregate easily because many of the other sediments contain a significant proportion of cement filled foraminifers and sand-sized clasts (Fig. 4-12A) identical to those in the larger clasts, while at the same time having numerous empty foraminifers (Fig. 4-12C).

The mineralogy and crystal habit of the cement, especially where easily visible in the *Globigerina*-rich grainstones, indicates precipitation from marine waters. The carbon and oxygen isotope ratios were determined for bulk samples of three intraclasts by R. M. Lloyd, Shell Development Company; δC^{13} ranges from +2·1 to +2·3 and the δO^{18} ranges from +2·2 to +2·9, versus the PDB standard. These analyses reflect the isotope ratios of the grains, matrix and cement together and suggest formation and cementation in the submarine environment. One intraclast was dated; it came from 105 cm in core 11 and is composed of *Globigerina*-rich mud. The radiocarbon age of this intraclast is 16, 535 ± 250 years B.P. Once again, while this may be the true radiocarbon age of sedimentation and lithification, because the rock is a mixture of pelagic sediment, mud and cement, the age is likely a relative one; the sediment may be older and the cement younger. The surface of this and other clasts is commonly veneered with serpulid worm tubes (Fig. 4-13A) and other encrusting biota, indicating that the fragments have had a complex and possibly long history, involving deposition, cementation, fragmentation, exposure on the sea floor and resedimentation.

South Water Cay (Fig. 4-14)

Both cores were taken in the distal fore-reef; core 9 (216 m) and core 11 (253 m). Most of the sediment is green-grey *Globigerina*-rich silt and mud, often containing up to 20% cement-filled pelagic foraminifers and sand-sized clasts. In core 11 these sediments are interrupted by segments of brown, intraclast-rich sediment that are sometimes in graded layers (Fig. 4-12B) or terrestrial silt. In the upper intraclast-rich layer (63–127 cm), the sediment is two-thirds granule-size and sand-size clasts and one-third *Globigerina*-rich silt. The lower layer (222–254 cm) is similar but also contains rare *Halimeda* plates and coral fragments, the only subsurface sediment to do so in either transect.

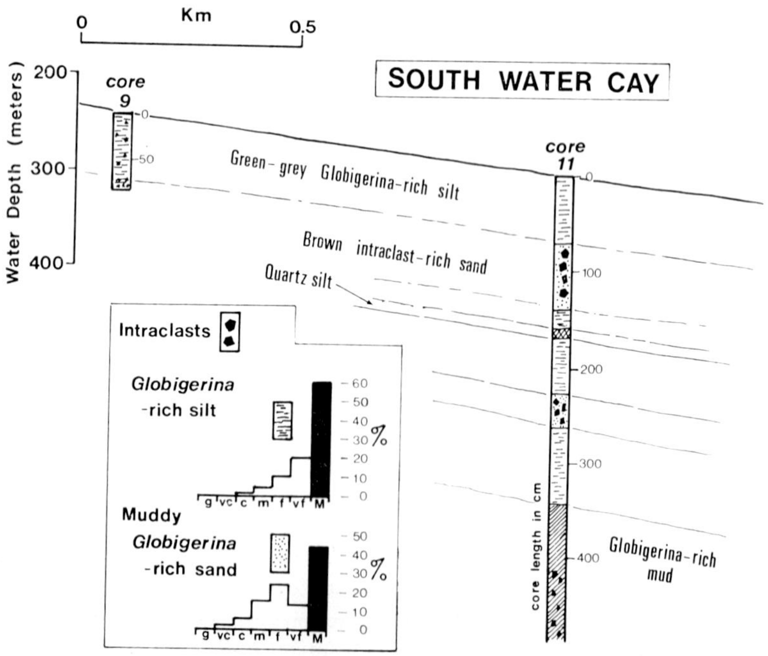

Fig. 4-14. A diagram showing the variation in sediment composition with depth in two sediment cores taken off South Water Cay (see Fig. 4-1 for location); note that the depth along the cores is in cm. In the boxed diagrams of grain-size distribution the upper plot is an average of seven samples from the uppermost unit and the lower plot is an average of six samples of the middle unit of core 11.

Tobacco Cay

The transition from distal fore-reef to basin is seen in three cores from Tobacco Cay (Fig. 4-15). Although the upper metre of core 4 (260 m) was lost, the lower part shows *Globigerina*-rich silt overlying an intraclast-rich layer 27 cm thick at 187 cm, near the same depth subsurface as that in core 11 off South Water Cay. This greenish-brown intraclast layer is a poorly sorted, muddy *Globigerina*-rich sand with almost one third of the sediment composed of clasts together with coral, mollusc, *Homotrema*, benthic foraminifer particles and silt-size carbonate grains. The upper 45 cm of core 6, some 1·5 km basinward of core 4, is composed of brown *Globigerina*-rich sand, sometimes in graded layers (Fig. 4-12B), in which over two-thirds of the grains are cement-filled pelagic foraminifers and clasts (Fig. 4–12D). The lower part of the core is the typical *Globigerina*-rich silt of the distal fore-reef. Core 7 (400 m) is *Globigerina*-rich mud, perennial sediment of the basin.

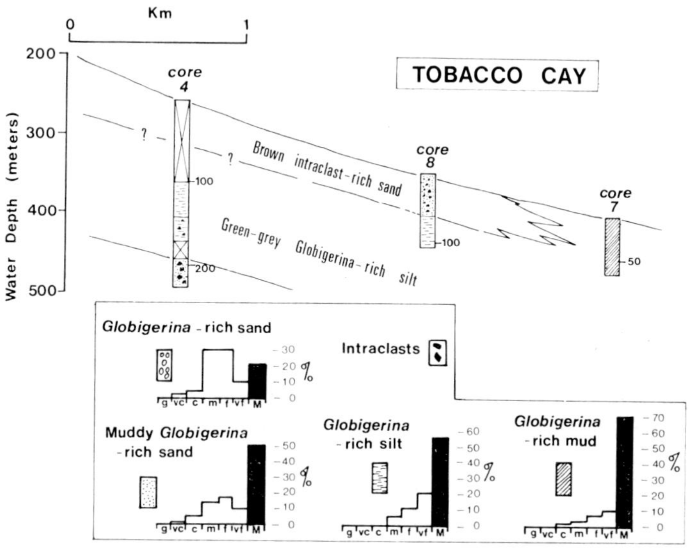

Fig. 4-15. A diagram showing the variation in sediment composition with depth in three piston cores taken off Tobacco Cay (see Fig. 4-1 for location); note that the depth along the cores is in cm. The grain-size plots in the box below the cores are for the different sediment types encountered.

Interpretation

The sedimentary sequence to a depth of about 5 m subsurface indicates that the style of sedimentation on the distal fore-reef and basin margin has been relatively constant. The layers of cemented intraclasts and terrigenous silt point to interruptions in this pattern.

The cemented intraclasts indicate that sediments are being lithified relatively rapidly by Mg-calcite on the distal fore-reef, either at the surface or just below. The fabric of the sediment, (broken and fragmented clasts together with skeletal particles, a mixture of cement filled and empty pelagic foraminifers and a grading of some sediment layers) indicates resedimentation. In this setting the source of such sediments must be upslope yet the composition of the sediments, with only rare *Halimeda* particles and coral grains, indicates that the source was not as high up on the slope as the proximal fore-reef and so must be local. We did not observe cemented hard-grounds on the distal fore-reef but seismic records indicate numerous faults that cut to the surface along the fore-reef between depths of 250 and 350 m (Fig. 2-4), the same depth as the clasts. We suggest that faulting has created small cliffs or steeply inclined segments that are free of modern sediment. The upfaulted sediments, exposed as cliffs, were lithified in a manner similar to that seen on the vertical walls of the Tongue Of The Ocean by Schlager & James (1978). Material that breaks off these cliffs episodically moves downslope.

The nearest present-day source for the layer of quartz silt in Core 11, Fig. 4-14, is the deep lagoon some 10–15 km landward of the barrier reef; it seems most improbable that quartz silt from the lagoon could have transported up and across the barrier under present conditions. It is more likely that the quartz silt was deposited during a glacial low stand when the present lagoon was exposed and streams carried siliceous sediment from the mainland across the barrier reef. On this interpretation, the layer of quartz silt might have been deposited during the last low stand of sea level between 15 000 and 80 000 years ago and therefore the upper 150 cm of sediment in these cores may be Holocene.

SEDIMENTS FROM DEEP PORTIONS OF THE CLIFFED FORE-REEF AND CAYMAN TROUGH

Observations from the submersible along that part of the southern barrier reef and Glovers Reef bordering the Cayman Trough revealed clear evidence of active sediment movement (see Chapter 3 p. 29–36). The upper part of this margin is precipitous and bathymetric profiles indicate that steep slopes extend to 1000 m and more. As we could observe only the upper 300 m, we sampled the deeper reaches of this part of the margin from shipboard to determine whether the episodic movement of upper margin sediments seaward makes a significant contribution to sediments in the Trough.

The steep margin of Glovers Reef and the southern barrier reef leads into two narrow deep elongate troughs on the western side of the Cayman Ridge (Fig. 4-16). Samples were taken along a transect seaward from Gladden Spit on the barrier reef and at three sites seaward of Glovers Reef (Fig. 4-16). Significant changes in sediment composition occur with depth along these two transects.

The shallow-water component of the sediments, reflected by the relative abundance of *Halimeda* plates, is variable. Off Glovers Reef sediments ponded on the slope are mixtures of *Halimeda* plates and pelagic foraminifers, to depths of 2000 m, some 30 km from the reef. In contrast, on the steep slope off the barrier reef sediments above 1000 m are *Halimeda*-rich skeletal sands and silts, as they are off Glovers Reef, yet below 1000 m they contain no *Halimeda* plates and are rich in pelagic foraminifers.

Above 1000 m in both areas sediments contain less than 10% insoluble residue. Below 1000 m the terrigenous component increases dramatically; off Glovers Reef mixed *Halimeda* and pelagic carbonates contain up to 30%; off the barrier reef there is progressive increase in insoluble residue with depth, 26% at 1100 m, 46% at 1600 m, over 60% at 2000 m, 84% at 2700 m. These sediments off Glovers Reef with a high insoluble residue are terrigenous muds with floating pelagic foraminifers. The deepest sediment, in the axis of one of the troughs that leads out of the Gulf of

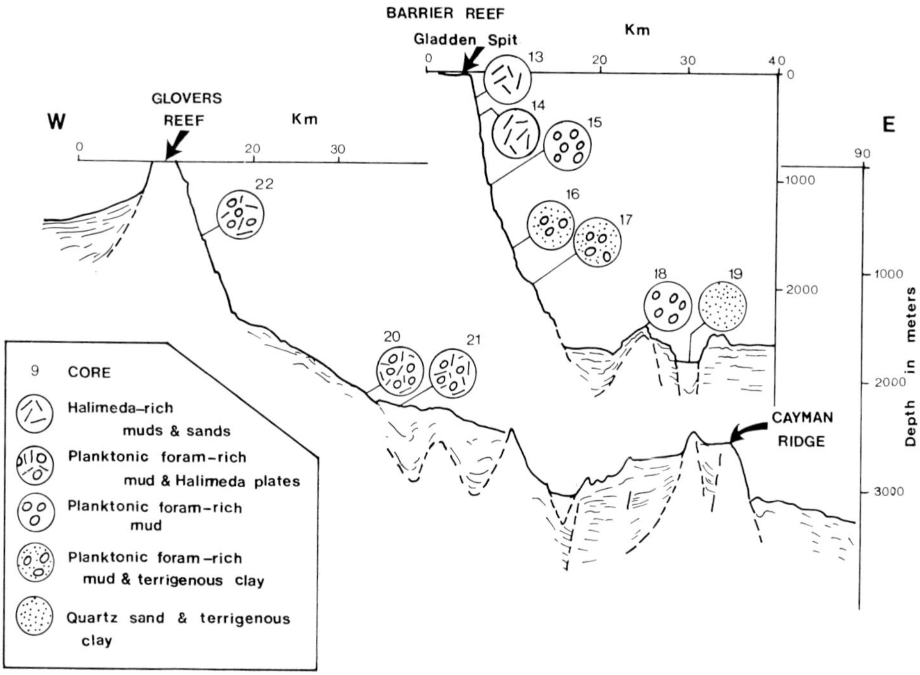

Fig. 4-16. The type of sediments collected from the top of short cores along the deeper reaches of the Belize continental margin; see Fig. 4-1 for locations.

Honduras (Figs 4-1; 4-16), is a medium-grained terrigenous sand with variable amounts of plant leaves and wood fragments along with pelagic foraminifers.

Tests of foraminifers and pteropods in sediments less than 600 m deep are often partially to completely filled with microcrystalline Mg-calcite cement (Brady, 1974). Tests in sediments of similar composition at depths of more than 600 m are empty.

These variations probably reflect several different factors, both physical and chemical.

The restriction of Mg-calcite precipitation to waters above 600 m is in agreement with the observation of Schlager & James (1978) who found that Mg-calcite precipitation for the large oceans of the world is localized above the perennial thermocline, around 1000 m.

Sediment is being transported into deep water (> 2000 m) as evidenced by the samples off Glovers Reef, yet this sediment movement is not ubiquitous, since there are no identifiable shallow-water components in deep water off Gladden Spit on the barrier reef. This apparent lack of movement of shallow-water sediment into deep water may reflect (1) the lack of sediment supply, (2) trapping of sediment high on the slope, or (3) dissolution of selected components with depth.

The terrigenous sediment found in the deep-water sediments may be a result of fluctuations in sea level during the Pleistocene and/or of dissolution of carbonate with depth. The red clay matrix between planktonic foraminifers on the steep slopes is possibly relict and was deposited in this region when sea level was lower during the Pleistocene and rivers draining the Maya Mountains flowed out onto the shelf. The zone of terrigenous mud, which now extends half way across the shelf, would extend out to the continental margin and through passes in the ancestral barrier reef. This clay would be mixed with the modern foraminifers by active bioturbation. The siliciclastic sand with terrestrial plant remains is likely derived from the Gulf of Honduras, the catch basin for rivers flowing out of the highlands of Guatemala and southern Maya Mountains, and was moved down the axis of the trough by turbidity currents and in suspension.

The role of dissolution is supported by the fact that in sediments below 1000 m, Mg-calcite comprises proportionally less of the total carbonate fraction with increasing depth; above 1000 m samples contain between 40 and 80% Mg-calcite, between 1000 and 2200 m they contain between 5 and 20% Mg-calcite, below 2300 m no Mg-calcite is found. The relative percentage of aragonite remains generally constant, between 25 and 50% throughout (M. Brady, personal communication).

SUMMARY

Sediments of the shallow reef are mostly coarse-grained coral–*Halimeda* lime conglomerates or lime sands with a grainstone fabric and little or no fine-grained sand. Deposits on the ledges of the wall are still coarse-grained sands but most of the very coarse-grained sediment is *Halimeda* plates. This same sediment is also found in between blocks and coral plates on the proximal part of the sloping fore-reef and contains less than 10% mud.

The first significant amounts of mud appear in sediments from the transition zone between proximal and distal parts of the sloping fore-reef, some 2 km away from the reef wall. Here the sediments are fine-grained, muddy, *Halimeda*-rich sands with a packstone to wackestone fabric. The *Halimeda* component in turn disappears from the sediments in the distal portion of the sloping fore-reef where sediments are mainly silts rich in planktonic foraminifers and have a wackestone fabric. In the basinal trough muds contain significant amounts of clay-size carbonate and planktonic foraminifers and are best described as pelagic carbonate muds with a wackestone fabric.

Internal sediments of the reef front and inside the wall are similar; both are composed of generally subequal proportions of granule, sand, and silt-size carbonate (little or no clay-size carbonate), and have a packstone fabric.

Sediments from the wall and fore-reef have three distinct modes in their grain-size distribution (granule-size, medium to fine sand-size, silt-size), and each mode has a different composition. The granule mode is almost exclusively *Halimeda* plates. The sand mode is predominantly fragments of molluscs, corals, coralline algae, benthic foraminifers and pelagic foraminifers. The silt mode is composed of skeletal material along with an abundance of sponge chips.

Shallow reef sediments can be distinguished from the wall and fore-reef sediments by the larger amounts of coral, but more importantly by the species of *Halimeda* in the granule-size fraction. In addition the fragments of the foraminifer *Homotrema* (an encrusting form), echinoid spines (particularly of *Diadema*) and articulated coralline algae (which grow best in agitated water) are characteristic of the sand-size fraction. The deeper water, perireefal sediments, besides lacking the above sand-sized grains in abundance, are composed primarily of *Halimeda* species that grow on the deeper parts of the reef.

The influence of the reef and wall as a source of sediment does not extend far into the basin; no reef-derived particles can be recognized in sediments only 4·2 km basinward from the wall.

The sediment of the distal fore-reef is *Globigerina*-rich skeletal silt with scattered intraclasts. The restricted occurrence of these intraclasts, the presence of an encrusting biota on some of them, their carbon and oxygen isotope ratios and the radiocarbon age of one specimen all indicate that they formed during the late Pleistocene on the sea floor. We suggest that they probably formed as thin surface hardgrounds or as the faces of small cliffs created by synsedimentary high-angle faulting on the distal fore-reef. The cemented sediments were subsequently disaggregated by burrowing or broke off the cliffs to produce the clasts.

The distribution of sediments in very deep water along the western margin of the Cayman Trough, which forms the edge of the southern part of the barrier reef and Glovers Reef, is different in the two areas studied. Off Glovers Reef *Halimeda*-rich sands with up to 30% insoluble residue occur to depths of 2200 m, some 30 km from the reef. Sediments off the barrier reef to the south, however, do not contain *Halimeda* or other shallow-water components below a depth of 1000 m. In addition these sediments contain progressively less carbonate with increasing depth below 1000 m. Precipitation of Mg-calcite subsea cement in the chambers of pelagic foraminifers appears to be restricted to depths less than 600 m.

Chapter 5

The composition and age of limestones from the reef front, wall and fore-reef

INTRODUCTION

A principal objective of our exploration of the reef margins in Belize was to sample the rock of the reef walls. The existence of steep slopes often with 'submarine cliffs' around oceanic reefs was well established by the time Darwin's classic work appeared in 1842. The rock forming these slopes, however, has rarely been sampled because of the danger of bringing a ship so close to shallow reefs and because of the difficulty of sampling such steep rocky slopes. A notable nineteenth-century exception is the work of T. Edgeworth David and his associates (David, Halligan & Finckh, 1904) who used a heavy chisel and hemp tangles from a small boat to collect samples down to 400 m off Funafuti Atoll.

To obtain samples of the wall and blocks on the fore-reef slope in Belize we used small charges of explosives implanted by the submersible. This method proved quite successful and allowed us to sample to depths of 174 m and to observe the internal structure of the limestone in the artificial outcrops.

This chapter gives descriptions of the artificial outcrops, the lithologies of the limestones and their radiocarbon ages.

LOCATION OF SAMPLES

Samples were collected by blasting at four of the seven areas studied: two are along the barrier reef, Tobacco Cay and Queen Cays (Figs 5-1, 5-3); the other two are around Glovers Reef off the seaward eastern side and off the leeward south-western side (Figs 5-2, 5-3).

The most intensively sampled locality is the barrier margin off Tobacco Cay where 5 sites were probed (Fig. 5–1). The top of the wall was sampled at 67 m (72-20), and the wall proper was sampled at three depths, 88 m (71-205), 97·5 m (71-207), and 110 m (72-22). The lower site was excavated twice (72-24) and sampled repeatedly. The large talus block on the fore-reef at a depth of 143 m was also sampled by blowing off one of the protruding corners (72-23).

The profile off Queen Cays is unlike the rest of the profiles along the barrier reef in that it does not flank the broad trough, instead the slope is precipitous into the Cayman Trough and the shallow reef is discontinuous. The wall here was sampled at a depth of 122 m with one charge (72-29) (Fig. 5-3).

The leeward side of Glovers Reef was sampled at one locality along the wall at a depth of 105 m (71-208) (Fig. 5-2).

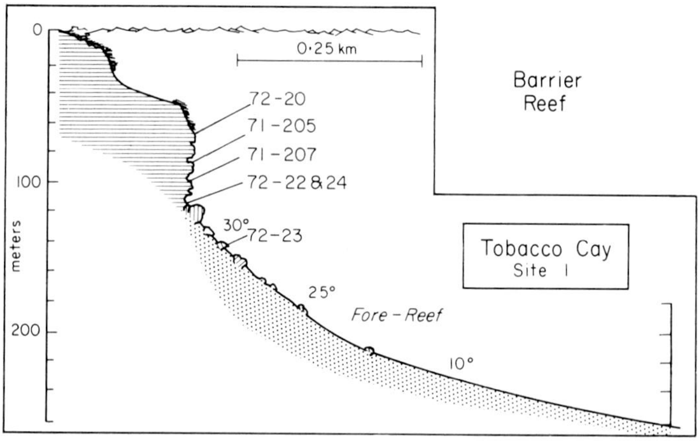

Fig. 5-1. A diagram of the shallow reef to basin transition off Tobacco Cay along the barrier reef (see Fig. 3-1 for location) with the rock sample locations indicated by numbers and arrows; see Table 5-1 for sample information.

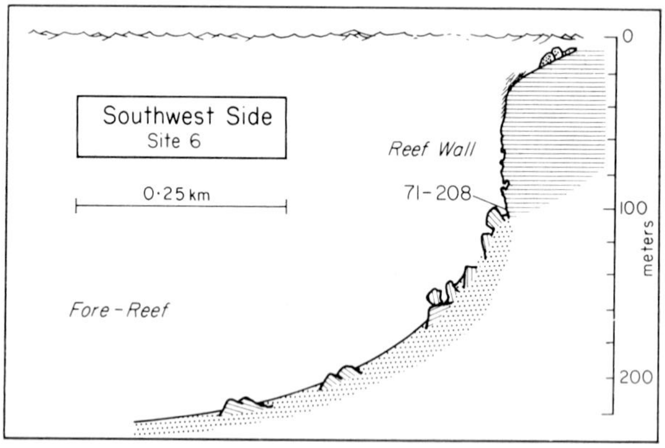

Fig. 5-2. A diagram of the shallow reef to basin transition off the southwestern side of Glovers Reef (see Fig. 3-1 for location) with the rock sample locality highlighted number and (arrow); see Table 5-1 for sample information.

The windward, eastern side of Glovers Reef was sampled at three sites: the step at 40 m (72-26 and 72-31); the base of the wall at 125 m (72-25); and the ridge and furrow structure at 174 m (72-27) (Fig. 5-3).

METHODS OF SAMPLING

To collect samples of limestone and attached organisms we used small charges of explosives implanted by the submersible. The explosive charges consisted of one-pound high explosive primers surrounded by up to ten pounds of Tovex Extra*, an explosive especially prepared to detonate at depth. A standard blasting cap was inserted in the primer and connected to the surface with heavy duty Primacord. The charge, explosive and primer were put in a one-gallon, wide-mounted plastic jar to which a metre-long wooden handle was affixed with tape. The prepared charge was

* Du Pont de Nemours, Wilmington, Delaware.

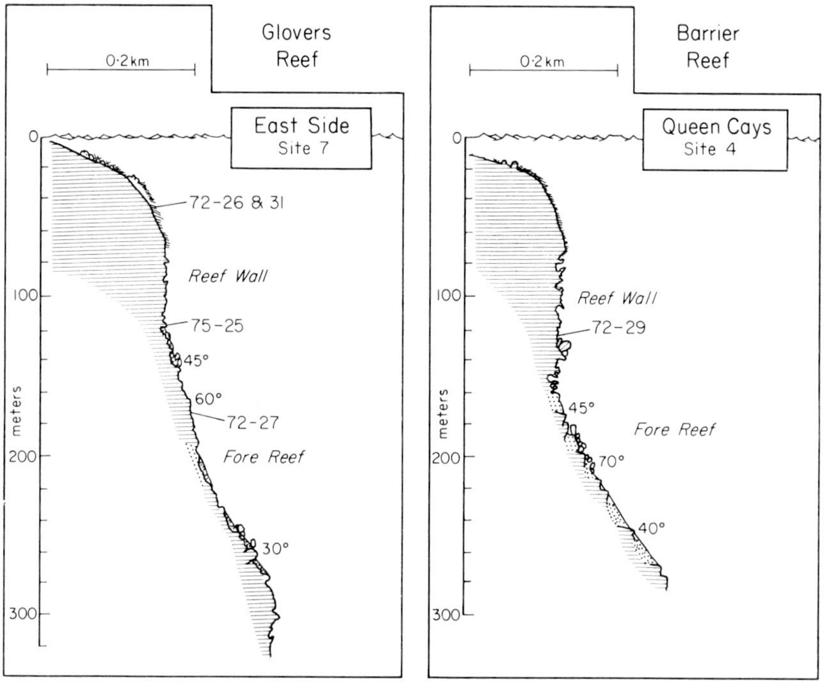

Fig. 5-3. A diagram of the upper 300 m of the western margin of the Cayman Trough at Queen Cays along the barrier reef and on the eastern side of Glovers Reef (see Fig. 3-1 for location) with the rock sample localities highlighted (numbers and arrows); see Table 5-1 for sample information.

carried by grasping the wooden handle firmly in NEKTON's mechanical claw. As the submersible descended with the charge, the Primacord, taped to light manilla line to prevent kinking, was played out from the support ship. When NEKTON reached the pre-selected location, the charge was positioned, usually in a cave or re-entrant of the rocky slopes. Once the submersible had returned to the surface and the hatch was opened, the Primacord was ignited with an electrical blasting cap attached by a buoy that could be floated clear of the support ship.

Within an hour or less after the explosive charge was set off the water was clear enough to return to the blast site, collect specimens of the rock, and photograph the artificial outcrop. Specimens weighing up to 10 kg were picked up from the rubble or on occasion picked directly from the exposure. Up to about 50 kg of specimens could be brought up from each dive. The location, size and amounts of samples collected from each site are given in Table 5–1.

STEP

Artificial exposure

The reef front was sampled at a depth of 40 m on the eastern side of Glovers Reef (Fig. 5-3 for location). Two successive charges produced a cavity about 7 m wide and extending an estimated 5 m into the wall (Fig. 5-4A). The site was accessible by SCUBA.

Table 5-1

Area		Sample Site No.	Depth (m)	Samples collected Weight (Kg)	%	Samples
Reef Front						
Glovers Reef	—windward side	72–26	40·0	40·8	8·9	38
		72–31	40·0	32·4	7·1	17
Top of wall						
Barrier reef	—Tobacco Cay	72–20	67·0	18·3	4·0	25
The wall						
	—Tobacco Cay	71–205	88·0	10·8	2·4	7
	—Tobacco Cay	71–207	97·5	9·0	2·0	10
	—Tobacco Cay	72–22	110·0	13·0	2·8	15
	—Tobacco Cay	72–24	110·0	136·8	30·1	81
	—Queen Cays	72–29	122·0	27·0	5·9	21
The wall						
Glovers Reef	—leeward side	71–208	105·0	10·3	2·3	8
Glovers Reef	—windward side	72–25	125·0	32·4	7·1	25
Sloping fore-reef-barrier						
	—Tobacco Cay	72–23	143·0	34·8	7·6	14
Cliffed fore-reef						
Glovers Reef	—windward side	72–27	174·0	90·0	19·8	80
	Total			455·6	100	341

Fig. 5-4. Site 1, depth 40 m: a cavity, about 7 m across, excavated into the face of the step.

Swimming into the cavity, one saw walls composed almost entirely of *Montastraea annularis* piled one on top of another with occasional colonies of *M. cavernosa* and *Agaricia* sp., all imbedded in soft sediment (Fig. 5-4B); the *M. annularis* are mainly broad and flat and many are upright in growth position. When entire colonies were picked from the walls fine-grained sediment flowed out of the face (Fig. 5-4C). By the time four or five specimens were collected from the far side of the cavity the water was so clouded with suspended sediment that subsequent samples had to be collected by feel. Swimming over the crest of the step, we saw air bubbles from divers exhalations in the cavity streaming out of the surface above the cavity, an indication of the high permeability of the 8–10 m of coral and soft sediment.

All the corals seen and sampled in the exposure are platey; individual colonies range from 1–10 cm thick with thinner plates of *Agaricia* most common near the surface; one series of overlapping plates fused to each other and more than a metre across was seen. In addition to the two species of *Montastraea* spp. and *Agaricia* spp., there were individual colonies of *Porites astreoides, Siderastrea, Colpophyllia natans* and *Scolymia cubensis* (Table 5-2).

Table 5-2. Number of specimens of various coral species in limestone collected from artificial exposures

Area	Site	*A. palmata*	*A. cervicornis*	*P. porites*	*P. astreoides*	*Siderastrea*	*Diploria*	*M. annularis*	*M. cavernosa*	*Agaricia*	*Mycetophyllia*	*Colpophyllia*	Total
Reef Front	72–26					1		20	1	5		1	28
	72–31			1				10		2			13
Wall	72–20					1		20					21
	71–205												0
	71–207				1	1		1					3
	72–22			1	1				2	1			5
	72–24		3	1	1	1		17		7	2	1	33
	72–29							3	1	3		1	8
	71–208							3				1	4
	72–25		1			1		5	3		1		11
Talus block	72–23							7		2			9
	72–27	1	3		1	2		8	4	9		2	30
Total		1	7	3	4	7		94	11	29	3	6	165

Only one sclerosponge was recovered, *Ceratoporella nicholsonii*, and it was alive, growing on the underside of a small ledge composed of very altered coral plates and coralline algae (Fig. 5-5A).

Composition of limestone

Cemented outer rind

The outer metre of the exposure is either interlayered plates of *Agaricia* and crustose coralline algae, or coral plates alone surrounded by well cemented, mottled *Halimeda* calcarenite (Fig. 5-5A) that shows multiple-generations of sediment-filled borings (Fig. 5-5B). The interlayered coral and crustose algae (Fig. 5-5A) is the

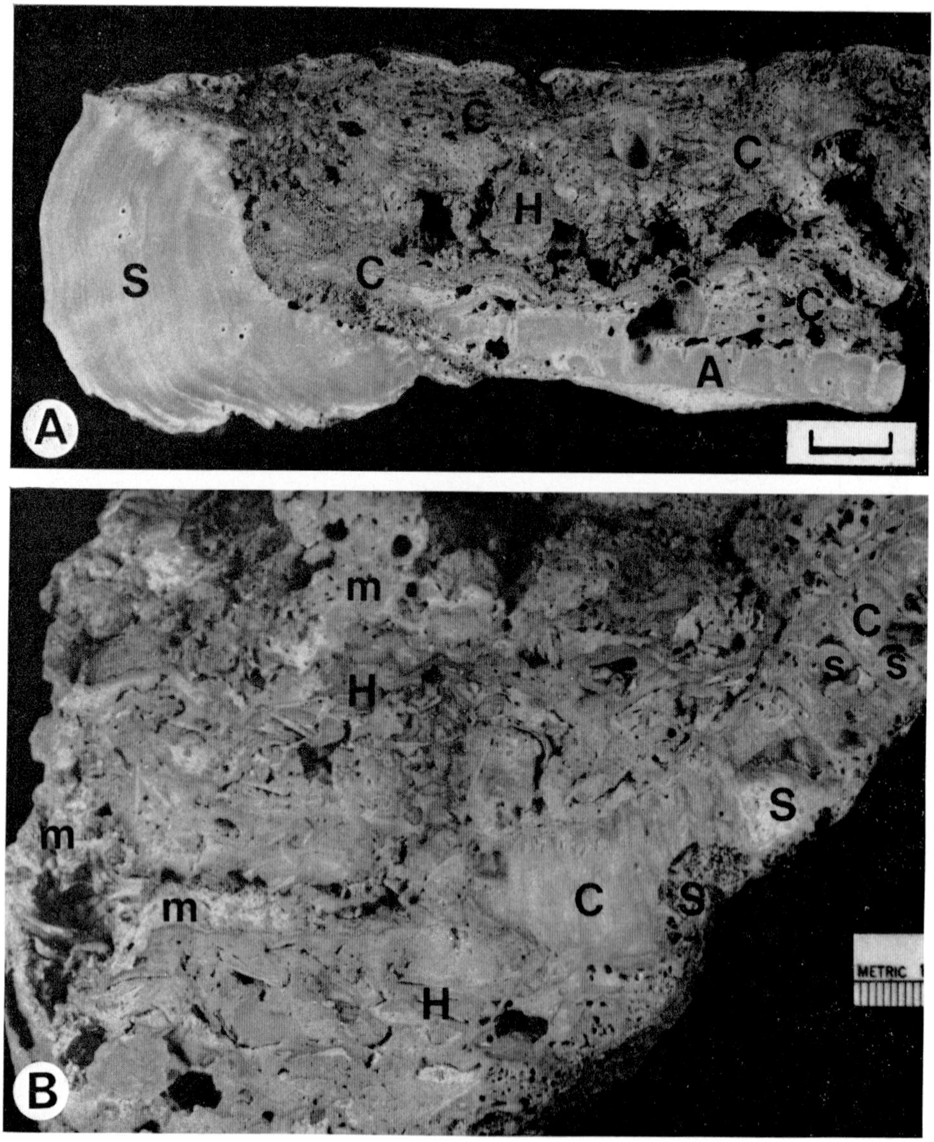

Fig. 5-5. Site 7, depth 40 m: (A) A polished slab of limestone from the well cemented outer zone of the sampling cavity illustrated in Fig. 5-4. The plate-like *Agaricia* sp. colony at the base [A] is overlain by an intergrowth of crustose coralline algae [C] and *Halimeda* wackestone [H] which has been bored by an endolithic sponge creating many cavities. The outer portion of this ledge is encrusted by a sclerosponge (*Ceratoporella nicholsonii*) [S]; scale bar 2·0 cm. (B) A polished slab of limestone from the same locality as Fig. 5-4. The rock is composed of a plate-like colony of *Montastraea annularis* [C] and multigeneration *Halimeda*-rich wackestone to packstone. The coral has been bored by sponges and the cavities are now partially infilled with sediment, mostly mud [S]. The sediment adjacent to the coral [H] has also been bored after lithification and the cavities are now filled with a second generation of fine-grained sand and mud [m]; scale in mm.

preserved equivalent of the numerous ledges of living coral projecting from the sloping surface here. Small cavities between the algal crusts are often filled with fine-grained cemented sediment, and the aggregate of corals, algae and cemented sediment shows multiple generations of borings filled with internal sediment, now lithified (Fig. 5-5B).

Unlithified interior

Corals beneath the outer metre commonly have encrustations of cemented sediment up to 2 cm thick on their topsides (Fig. 5-6A, B) and the bottoms may be bored or encrusted with serpulid worm tubes and foraminifers. Thin crusts of cemented mud on the upper surface of many other corals are either: (1) bumpy surfaces with micro-relief of up to a centimetre (formed by irregular pits and knobby projections (Fig. 5-6A), some of which have twig-like growths on their apices); or (2) smooth mammillary surfaces composed of laminated to mottled lithified lime mud that drapes over many irregular surfaces to give the top of the coral a smooth appearance. These smooth or bumpy crusts up to 2 cm thick show both sharp and gradational contacts with the underlying coral skeleton (Fig. 5-6B). These are similar to crusts described by Land & Goreau (1970) from Jamaica and Macintyre (1977) from Panama.

The inter-coral sediment is a mixture of *Halimeda* plates and lime mud that is mostly uncemented. Here and there, often around the coral plates, the sediment is a lightly cemented meshwork of stacked and interleaved *Halimeda* plates with fine-grained sediment between the plates (Fig. 5-6C). The resulting friable rock is easily disaggregated by handling and it comprises no more than 5% of the total rock specimens recovered.

Radiocarbon ages

A *Montastraea annularis* colony from the outer 2 m of the limestone, yet beneath the well cemented outer rind, has a radiocarbon age of 4985 ± 110 years B.P. (Table 5-3). Another *M. annularis* colony collected from the same unlithified material, but 4 m in from the surface of the slope (72-31), has radiocarbon age of 5365 ± 140 years B.P. The radiocarbon age of fine, unlithified muddy sand perched on top of the cup-like *M. annularis* colonies from the inner reaches of the cavity (72-31) is 4970 ± 75 years B.P. These concordant ages indicate that both the corals and sediment of the outer 4 m were deposited together some 4000 to 5000 years B.P.

THE WALL AND FORE-REEF

The wall and its extension were sampled at ten different sites; in addition, samples were collected from a large talus block on the fore-reef slope.

Almost all the samples from the ten sites are well cemented mixtures of varying proportions of reef-building corals, *Halimeda* plates, skeletal sand, and mud. Most of the specimens with appreciable sediment have a characteristic, highly irregular mottling produced by variations in grain-size and composition on the scale of centimetres. This mottling is the result of reiterated boring, filling of the bores with sediment and cementation of the sediment. The resulting multigeneration, complex limestones, with exception of specimens from one site, all have radiocarbon ages of less than 15 000 years B.P.

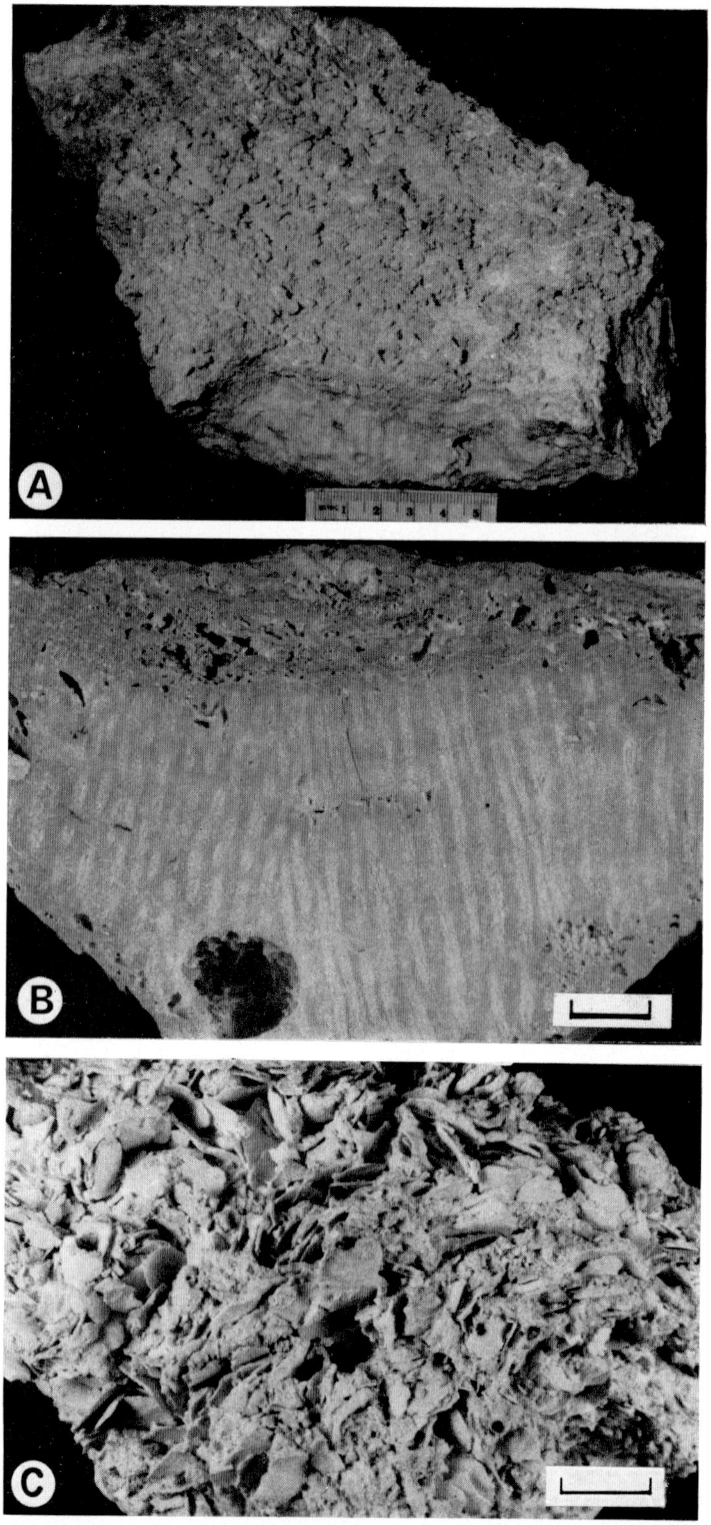
A
B
C

The significant features of the artificial exposures and the specimens are described in the following sections, followed by a summary of the principal lithologies and the radiocarbon ages.

TOP OF WALL

Artificial exposure

At the top of the wall off Tobacco Cay (Fig. 5-1) the bottom is made up of many mounds or hummocks 2–4 m across and several metres high, capped or completely covered with overlapping plates of coral, generally *Agaricia grahamae*, with bare rock or streams of sand winding down the slope and over the edge between hummocks. A charge was placed in a cavity beneath one of these hummocks at a depth of 67 m (72-20); it removed the hummock, shattered coral colonies on adjacent promontories (Fig. 5-7A), created a cavity about 6 m high and 4 m wide, and formed a concave scar that extended back into the rock wall 3–4 m.

The upper 3 m of the cavity is a series of slightly curved coral plates, mainly *Agaricia* with minor *Montastraea*, piled on top of one another (Fig. 5-7B), with lithified sediment between. The lower 3 m, partly obscured by loose sediment from above, is mostly *Halimeda* calcarenite, much of which appeared to be unlithified.

Composition of limestone

Over half of the material collected is coral plates, almost all *Agaricia* sp. with only two *M. annularis,* one *Colpophyllia natans,* and one *Montastraea cavernosa* colony. Surfaces of isolated corals and the corals surrounded by lithified sediment are encrusted with attached serpulid worm tubes, *Homotrema rubrum*, solitary corals such as *Scolymia cubensis,* a few bivalves, and worm tubes composed of well lithified fine sediment.

Individual coral plates are rarely more than 5 cm thick and most are less than 3 cm. The interiors of these thin plates are invariably bored, with cavities excavated by the sponges *Cliona* spp. and *Siphonodictyon* spp. These borings have in most cases removed more than half of the original skeleton. The holes are commonly filled with lithified sediment but may contain loose material as well. As a result of this extensive alteration, insufficient original coral aragonite was present in any of the samples for radiocarbon dating.

The remaining half of the material recovered is about equal proportions of *Halimeda* packstones to wackestones, and *Agaricia* plates with the same *Halimeda* packstone to wackestone between them.

The limestone with *Agaricia* plates is invariably much altered; both the lithified sediment and corals are riddled with cavities from sponges, worms, and other unidentified organisms, and the many generations of cavities are filled for the most part

Fig. 5-6. Site 7, depth 40 m: (A) A view of the topside of a broken slab of *Montastraea annularis* with a bumpy crust of cemented sediment from the inner, poorly lithified part of the cavity illustrated in Fig. 5-4; scale bar in cm. (B) A polished slab of a plate-like colony of *Montastraea annularis* with a bumpy crust of cemented sediment on the topside, from the inner poorly lithified part of the cavity. The sediment consists of peloid-rich silt cemented by Mg-calcite micrite; scale bar 1·0 cm. (C) A sample of friable *Halimeda* grainstone from between colonies of *Montastraea annularis* in the inner part of the cavity illustrated in Fig. 5-4; scale bar 2·0 cm.

Table 5-3. Radiometric age determination

Sample No.	Area	(Site)	Water depth (m)	Estimated distance behind rock face (m)	Material	^{14}C Age (Years B.P.)
	Reef step					
72-26	Glovers Reef	(7)	40·0	(2)	*M. annularis*	4985±110
72-31	Glovers Reef	(7)	40·0	(4)	*M. annularis*	5365±140
72-31	Glovers Reef	(7)	40·0	(4)	Unlithified sediment	4970±75
	Reef wall					
71-207.10	Tobacco Cay	(1)	97·5	(0)	*M. annularis*	2430±95
71-207.2	Tobacco Cay	(1)	97·5	(1)	*M. cavernosa*	7835±145
72-22a	Tobacco Cay	(1)	110·0	(1)	*C. natans*	8100±190
72-22.5a	Tobacco Cay	(1)	110·0	(1)	Friable white sediment	15 220±205
72-24	Tobacco Cay	(1)	110·0	(5)	*M. annularis*	9320±120
72-24.4a	Tobacco Cay	(1)	110·0	(5)	*M. annularis*	13 160±280
72-24.48	Tobacco Cay	(1)	110·0	(5)	*M. annularis*	13 195±215
72-24.16	Tobacco Cay	(1)	110·0	(5)	Well lithified *Halimeda* grainstone	7920±130
72-24.55	Tobacco Cay	(1)	110·0	(5)	Well lithified *Halimeda* packstone	9405±115
72-24.4b	Tobacco Cay	(1)	110·0	(5)	*Halimeda* packstone*	12 240±150
72-24.28	Tobacco Cay	(1)	110.0	(5)	*Halimeda* grainstone†	2235±70
72-24.3	Tobacco Cay	(1)	110·0	(5)	Well lithified laminated calcilutite	11 925±130
72-29	Queen Cays	(4)	122·0	(1)	*M. annularis*	12 747±190
72-25	Glovers Reef	(7)	125·0	(1)	*S. radians*	23 820±600
72-25.22b	Glovers Reef	(7)	125·0	(2)	*S. radians*	32 280±1620
	Talus block					
72-23	Tobacco Cay	(1)	143·0	—	*M. annularis*	7095±120
72-23	Tobacco Cay	(1)	143·0	—	Botryoidal aragonite	12 740±250
72-23	Tobacco Cay	(1)	143·0	—	Botryoidal aragonite	12 600±150
	Ridge and furrow					
72-27.1	Glovers Reef	(7)	174·0	(2)	*M. cavernosa*	7865±140
72-27.2	Glovers Reef	(7)	174·0	(2)	*A. cervicornis*	11 770±165
	Modern sediment					
72-30.2	Tobacco Cay	(1)	116·0	—	*Halimeda* sand	835±60

* Two generations of sediment infill.
† Last stage of sediment infill in a very altered rock, very friable.

with laminated mudstone and occasionally *Halimeda* grainstone. The one *Colpophyllia* is intensively bored, and the holes are filled with mudstone.

THE WALL

Artificial exposures

The wall between depths of 70 and 120 m was the zone most intensively sampled. Off Tobacco Cay samples were recovered from blast sites at 88, 97·5, and 110 m (Fig. 5-1) while off Queen Cays, also along the barrier reef, samples were recovered from 122 m (Fig. 5-3). Glovers Reef was sampled both on the leeward side at 105 m and on the windward side at 125 m (Fig. 5-2) and at 174 m (Fig. 5-3). Samples recovered from the blast sites range from a few centimetres in size to 70 cm long and 20 cm wide.

All of the limestones of the wall are brittle, as evidenced by the numerous cracks radiating out from the blast site; yet the rock is cavernous, with many irregular cavities ranging from a few centimetres to several metres wide. The limestone is about half corals; *M. annularis, Porites astreoides,* and *Mycetophyllia* in laminate to massive growth forms, both in growth position and overturned (Fig. 5-7C), were the most noticeable in the artificial exposures. Where visible, the matrix between corals is composed of *Halimeda*-rich limestones and many of the open cavities were lined with dark brown to black crusts.

Composition of limestones

The limestones from all these localities are remarkably alike: varying proportions of massive and platey corals and *Halimeda*-rich sediment, both of which are often altered by reiterated boring, sediment infill, and lithification. Samples from localities up to 30 km apart are so similar that they cannot be differentiated. Because the limestones are so similar they are described together.

Corals

The commonest elements are fragments, pieces, and almost whole colonies of massive corals. These colonies are usually clearly from large irregular to rounded head-like forms with rare lobate and rare platey forms. Although some individual corals were recovered, they were usually encrusted with coralline algae and clearly come from the outer surface of the cliff and not the main bulk of the rock behind.

The colonies are sometimes encrusted with coralline algae but most often the periphery of the skeleton is obscure and grades into lime mudstone, which in turn grades into the surrounding lithified sediment (Figs 5-8A, 5-9). The interior of many colonies is riddled with cavities, large and small, that are the result of boring by sponges and bivalves (Fig. 5-9), and with irregular cavities of uncertain origin (also see Chapter 6). The cavities are sometimes empty, but often are partially or completely filled with lithified sediment and/or cement.

The commonest massive coral is *Montastraea annularis* (Fig. 5-8), which was recovered from almost all wall sites and is the commonest coral in all but one (see Table 5-2). It comprises almost half of all the corals collected in samples from the wall. The other common massive corals are *Porites astreoides, Siderastrea* sp., and *Montastraea cavernosa* (Fig. 5-9). A frequent but not volumetrically significant coral is *Colpophyllia natans.*

Less common, although ubiquitous, are platey corals. These colonies occur in samples from the wall as individual plates with lithified sediment crusts on top, or as the roofs of cavities partially filled with lithified sediment (Fig. 5-10). Less commonly the plates are covered with only a thin layer of sediment, indicating that the cavity was not filled. Finally they are isolated plates that along with other corals float in lithified mud or *Halimeda* calcarenite (Fig. 5-11).

The commonest platey coral is *Agaricia* spp., and this is the most frequently

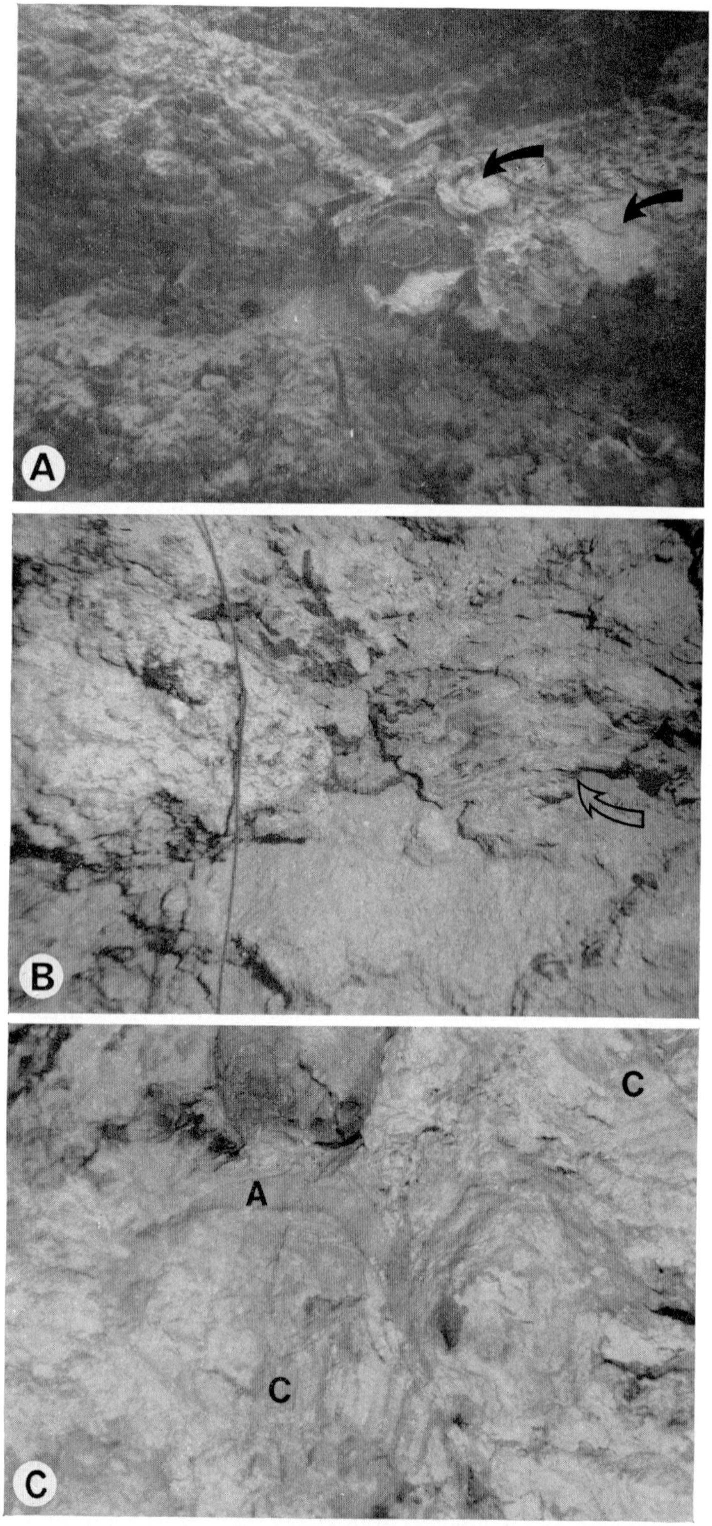
A
B
C
A
C
C

collected coral after *M. annularis*, although in these rocks it is volumetrically much less important that the massive corals (see Table 5-1). These thin, platey colonies are often bored by sponges prior to the infilling of surrounding sediment, as shown by the interior of the plates which are free of borings. The borings in coral plates are commonly filled with calcilutite and/or cement.

The branching forms, *Porites porites* and *Acropora cervicornis* are occasionally present, but are most often found in compound rocks composed of *Halimeda* calcarenites and pieces of coral (see below).

Squamariacean algae

In addition to the familiar crustose corallines there are also irregular sheety growths of *Squamariaceae,* a family of encrusting algae that secrete an aragonite skeleton (Wray, James & Ginsberg, 1975; Wray, 1977). They are both intergrown with corallines as encrustations on corals (Fig. 5-9) and separate from the corals and corallines as many arched and domed crusts that are rarely in contact with one another. The cavities between crusts, 5–10 mm apart, are commonly partially filled with sediment.

Sediments

In the specimens collected from the wall, well lithified sediments of widely varying grain size occur with and without branched, platey, and massive corals. The variations in grain size extend to the level of individual specimens in which there are irregularly shaped areas with sharp boundaries of two or more textures—mudstone, packstone, grainstone—and it is not uncommon to find all three textures in a single large specimen.

The wide variations in texture are paralleled by variations in structure; some grainstones and packstones are thin-bedded with layers 2–3 cm thick; others are homogeneous; still others are mottled as a result of smaller scale variations in texture. The mudstones commonly occur as laminated fillings of isolated voids a few centimetres in size within coral or cemented sediment; with alternately dark and light laminations up to a few millimetres thick.

A common denominator of all the different textural types is the presence of varying amounts of *Halimeda* plates or fragments of plates. The plates or fragments may be arranged horizontally in packstones and grainstones, they may be stacked together irregularly to form an open meshwork filled with finer sediment, or they may be floating in mudstone.

The shapes and generally sharp boundaries of the different textural types within individual specimens make it clear that they are internal sediments deposited in voids made for the most part by animals boring into coral or cemented sediment, but in

---

Fig. 5-7. Site 1, (A) depth 67 m: the crest of the wall off Tobacco Cay several metres away from the blast site. The mound (about 3 m across) capped with many overlapping plate-like colonies of *Montastraea* has been partly shattered and some of the colonies have broken away (light grey areas, arrow). (B) Depth 67 m: a close view of the rock exposed at sample location 72–20 composed of many overlapping plates of coral (arrow); compare to the structure of the mound illustrated in (A); the basting wire at left is about 0·5 cm in diameter. (C) depth 110 m: the limestone exposed by an excavation into the wall. The rock is composed of reef corals such as *Montastraea annularis* [A] and *Montastraea cavernosa* [C] with the sediment between them sand and silt. The field of view is about 0·7 m across.

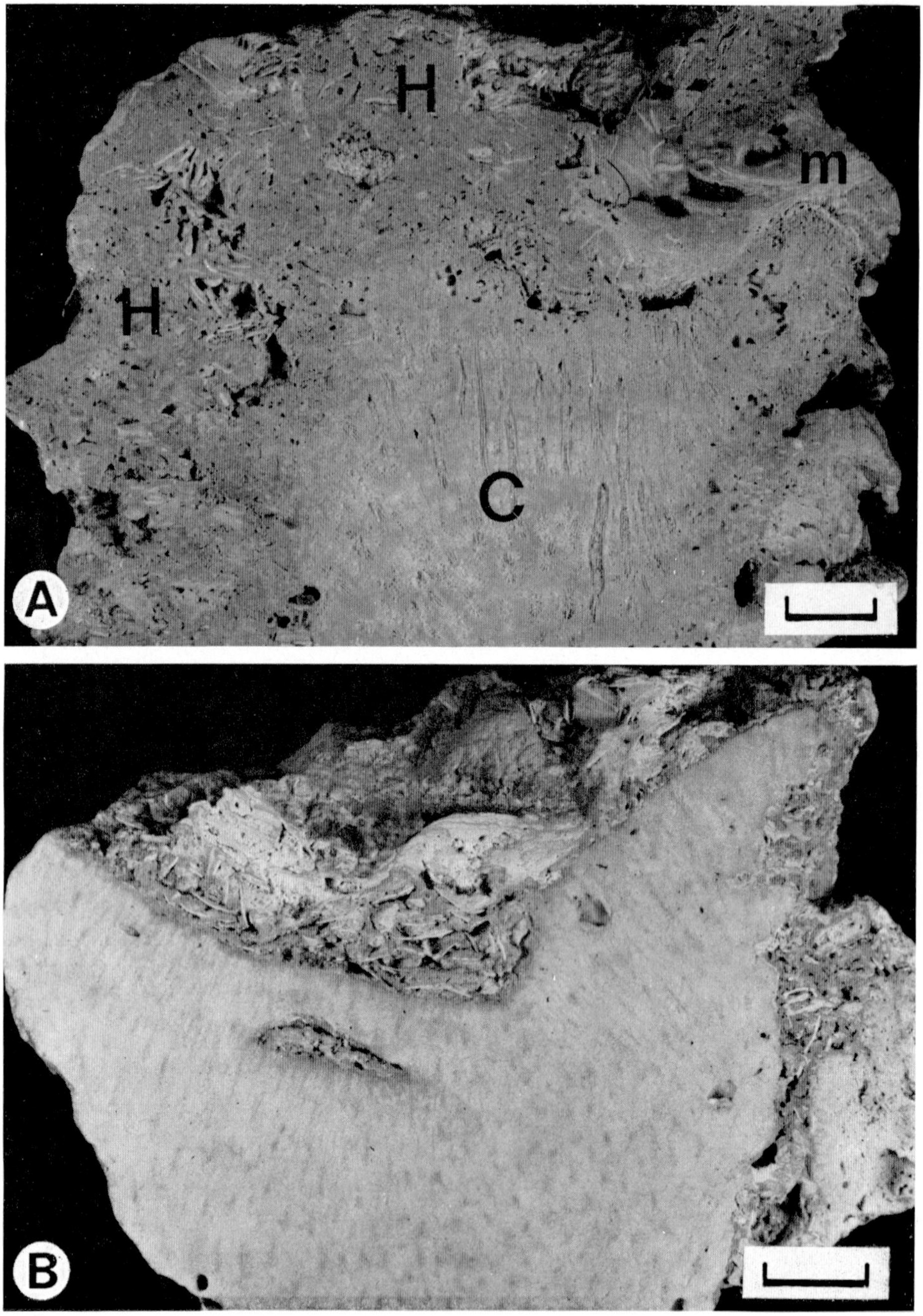

Fig. 5-8. Site 1, depth 110 m: (A) a polished slab of *Montastraea annularis* [C] and surrounding sediment. The diffuse boundary between the coral and sediment is produced by multigeneration boring, sediment infill and early cementation. The sediment [H] is a *Halimeda* packstone to wackestone with a second generation of mud [M] infill in the upper right. Shelter voids are produced by the *Halimeda* plates in the sediment; scale bar 2·0 cm. (B) A polished slab of *Montastraea annularis* and surrounding *Halimeda* grainstone to mudstone. The *Halimeda* plates form numerous shelter voids while the overlying mud contains numerous drapes and small discontinuities; scale bar 2·0 cm.

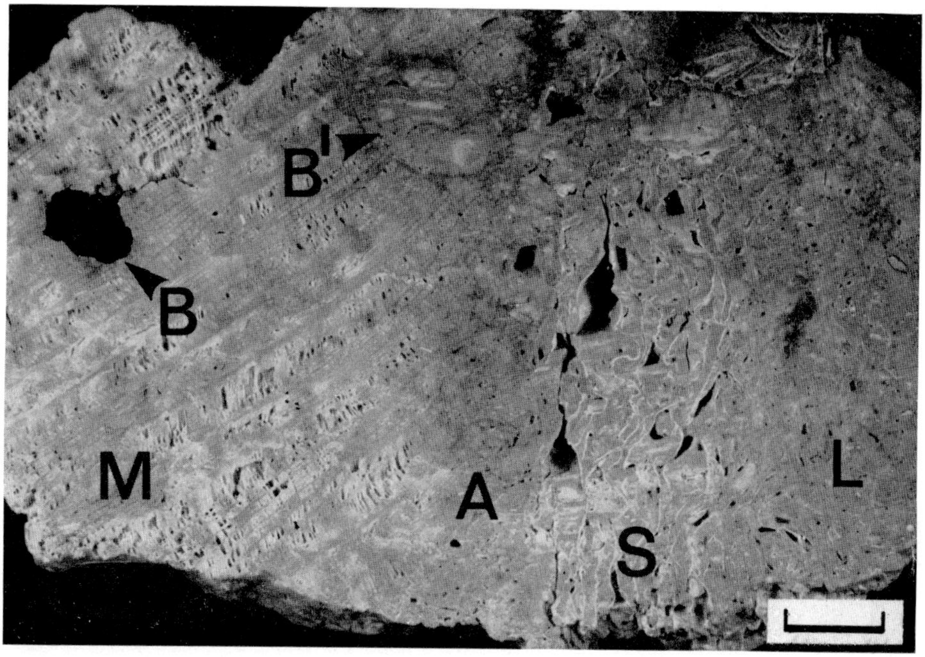

Fig. 5-9. Site 1, depth 110 m: a polished slab of *Montastraea cavernosa* (M) partially bored (B) with an altered periphery (A) and encrusted with squamariacean algae (S) which is buried by mudstone (L). The large ovoid mollusc boring in the coral (B[1]) is filled with laminated sediment. The micromottling of A is produced by many generations of boring, sediment infill and lithification. The layers of the squamariacean algae are the white leaves that grew free, not as encrustations, and the spaces between them were subsequently filled with mud; scale bar 2 cm.

some instances by burrowers into soft sediments. Successive generations of internal sediments, each of which is slightly different, give the wall limestones their characteristic texture mottling.

Radiocarbon ages

The radiocarbon ages of corals and lithified sediments from the wall are given in Table 5–3: six specimens of coral gave ages between 7800 and 13 200 years B.P.; five samples of cemented sediment, grains and cement, have a similar range of ages, 7900–15 000 years B.P. For one specimen (72-24·4, Table 5-3), comparison of the age of a coral, 13 160 ± 280 years B.P. with that of the cemented sediment surrounding it, 12 240 ± 150 years B.P. suggests that growth of the corals, formation of the sediment and cementation all took place at about the same time.

Only four samples have radiocarbon ages that fall outside the range 7900–15 000 years B.P. Two are corals from the base of the wall off Glovers Reef with apparent* ages of 23 000 and 32 000 years B.P. Of the two samples with younger ages, one is an intensively bored and encrusted coral from the surface of the wall (2430 years B.P.) and the other is a very friable, late stage cavity-filling sediment (2235 years B.P.). Because surface sediment has a radiocarbon age of 835 years B.P. (Table 5-3), these

* Because of the low levels of ^{14}C, these ages are suspect and the rocks may be much older.

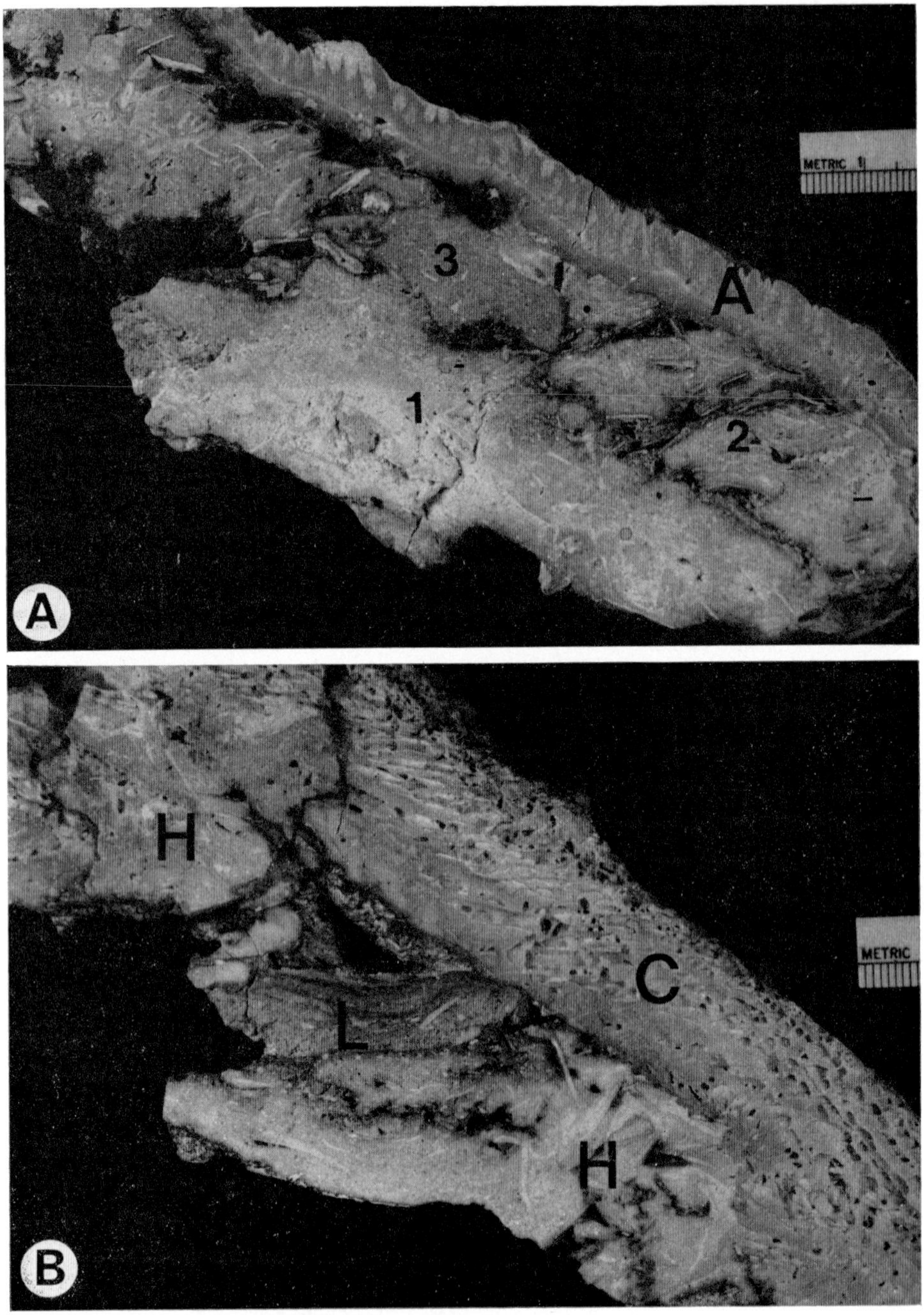

Fig. 5-10. Site 1, depth 110 m: (A) a polished slab of coral and lithified sediment. The coral, a plate-like colony of *Agaricia* sp. [A] has acted as a roof with the sediment in the cavity beneath deposited as crudely layered internal sediment, most of which is *Halimeda* grainstone to packstone. Generations of sediment infill (1, 2, 3) are separated by iron-oxide coatings. Scale in cm. (B) A polished slab of coral [C] (*Mycetophyllia* sp.) which has acted as a roof and two generations of sediment deposited beneath. The first generation is *Halimeda* grainstone [H] which was bored and the cavity filled with *Halimeda* grainstone-wackestone [L]; scale in mm.

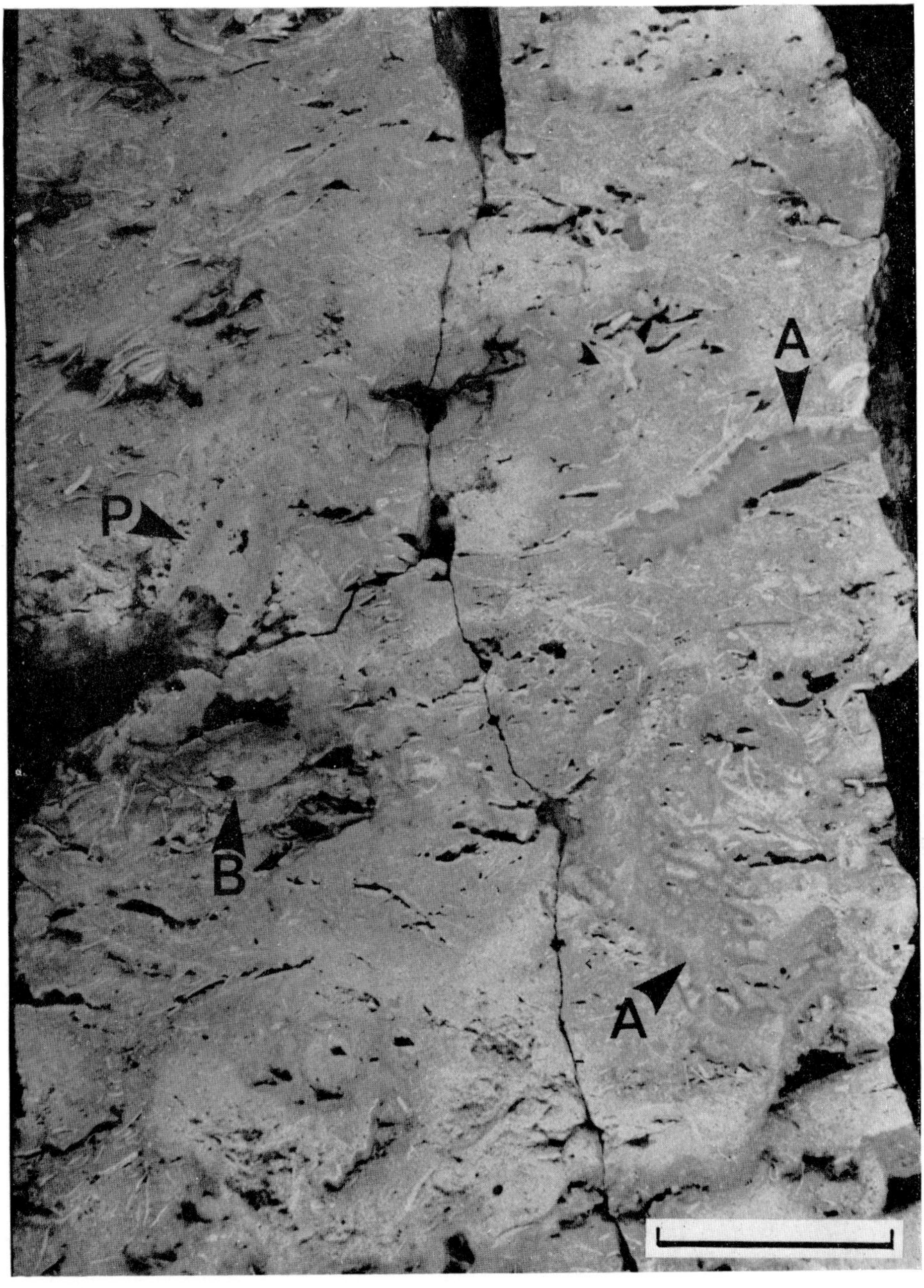

Fig. 5-11. Site 1, depth 110 m: a large slab of wall limestone composed of coral and *Halimeda*-rich sediments. The two corals in this sample are *Agaricia* sp. (A) and *Porites porites* (P). The sediment is mainly *Halimeda*-rich grainstone and packstone in which the *Halimeda* plates have created many shelter cavities. Several borings (one above the scale bar, the other at B) are partially filled with fine-grained geopetal sediment; scale bar 4·0 cm.

young ages indicate that sedimentation and cementation are not restricted to the period 7900–15 000 years B.P. The lack of specimens with ages between 7900 and 2400 years B.P., however, requires explanation.

LIMESTONE BLOCK ON THE SLOPING FORE-REEF

A large talus block, about 5–7 m on a side and resting on the sediment slope at a depth of about 143 m off Tobacco Cay, was sampled by blowing off about 2 m^2 of one corner (Fig. 5-1). The artificial exposure and samples reveal a hard and cavernous limestone identical in composition to that seen in exposures of the lower wall. A columnar colony on *M. annularis* in the limestone has a radiocarbon age of 7095 ± 120 years B.P., confirming that the limestone has the same apparent age as that of the wall limestone. Two samples of botryoidal aragonite cement from the block have radiocarbon ages of 12 600–12 740 years B.P., within the age range of samples from the adjacent Holocene wall limestone (see Table 5-2).

CLIFFED FORE-REEF

Composition of limestone

This deepest site sampled is located on one of the prominent ridges running down the slope on the eastern (windward) side of Glovers Reef at a depth of 174 m (Fig. 5-2). The topography of the ridge is a series of slightly overlapping ledges with smooth rounded surfaces and cavities between. Samples from this locality are characterized by their smooth scalloped exteriors, like the smooth walls of the cavities observed in the rock. These surfaces are well bored and encrusted with many epibionts conspicuous among which are solitary corals and lithistid sponges. The rock is composed of massive corals, platey corals, platey corals with sediment sandwiched between, and lithified *Halimeda* calcarenite. More than half of the samples are intensively altered by repeated generations of boring and sediment infill, with individual stages often separated by iron–manganese coatings. Much of the material is reminiscent of the samples collected at the break in slope at 67 m off Tobacco Cay.

Individual massive corals or large colonies (Table 5-2) make up about a third of the samples collected. Platey corals such as *Agaricia* and *Colpophyllia* are generally in the form of parallel colonies with lithified sediment, mainly *Halimeda* calcarenite sandwiched between (Fig. 5-12A, B). Particularly conspicuous at this site were colonies of *Acropora palmata* (Fig. 5-12C) and *Acropora cervicornis*. The remaining third of the rock is composed of *Halimeda* grainstone with scattered coral plates and calcilutite. These sediments are intensively bored, and the cavities infilled with later generations of *Halimeda* grainstone and calcilutite (Fig. 5-12B). One style of cemented sediment, noted only in the shallow blast site at 40 m above, is coral plates and sticks of *A. cervicornis* capped with a pile of cemented calcilutite.

Radiocarbon age

The radiocarbon age of a large *M. cavernosa* colony is 7856 ± 140 years B.P., and the age of an *A. cervicornis* stick cemented into the rock is 11 770 ± 165 years B.P., see Table 5-3. These samples are clearly much younger than the rock which makes up the base of the escarpment (23–32 000 years B.P.) some 50 m above.

SUMMARY

Limestone composition

The limestone that forms the outer several metres of the reef margin to a depth of almost 200 m is the same throughout with only slight differences in composition and cementation from shallow water to deep water.

The shallowest area sampled, the step off the seaward, eastern side of Glovers Reef, is characterized by the relative lack of lithified material. Most of the outer 4 m of the steep slope at a depth of 40 m is composed of numerous large, in place colonies of *M. annularis*, with the voids between partially to completely filled with mud and sand, only a minor part of which is even poorly cemented. The only cemented material is friable *Halimeda* grainstone and thin, lithified mudstone crusts on top of coral colonies. The outer metre or so at this locality appears to be a rind of well lithified coral and sediment, mostly plate-like colonies of *Agaricia* and *Montastraea*, with intervening *Halimeda* calcarenite and calcilutite that is extensively altered by several generations of cavity formation, sediment infill and lithification.

The relative lack of cementation in this shallow site is similar to that in the shallow reef above (James *et al.*, 1976) where to water depths of 10 m along both the barrier reef and the seaward side of Glovers Reef little of the outer few metres of the reef or surrounding material is thoroughly lithified.

The contact between the brow and wall sampled at a depth of 67 m off Tobacco Cay along the barrier reef, is composed largely of many plate-like coral colonies, mostly *Agaricia* species, between which is well cemented *Halimeda* packstone to grainstone. This rock is what would be expected from the living corals and the type of sediment that accumulates on the present day brow and wall at this depth. Much of the sediment here is also unlithified but large pieces of thoroughly lithified *Halimeda* packstone were also recovered. Most of the samples indicated extensive alteration by reiterated cavity formation, sediment infill, and lithification.

Most of the samples from the six sites on the wall and from the talus block have the same composition: corals such as *M. annularis,* with accessory *M. cavernosa, P. astreoides, S. radians,* and *C. natans* in rounded and massive growth forms characteristic of these species when growing in shallow water, surrounded by well lithified *Halimeda* packstone to grainstone and laminated to mottled mudstone. Less common, although present, are plate-like corals such as *Agaricia* forming the sides of shelter cavities or, if small, forming part of the sediment between larger massive forms. Stick-like colonies of *P. porites* and *A. cervicornis* are also present as clasts between larger colonies. The coral and lithified sediment occur in about equal proportions. About half of the rocks are altered by cavity formation, sediment infill, and lithification, and in some cases as many as five generations of the process can be recognized in a single specimen. Cavities are sometimes partially to completely filled with mammilary botryoidal aragonite (Ginsburg & James, 1976).

At the deepest locality sampled, the deep ridge and furrow structure of the cliffed fore-reef on the eastern side of Glovers Reef at a depth of 174 m, the limestone is slightly different from that of the wall; it is composed of massive corals and intervening *Halimeda*-rich calcarenite, but the proportion of plate-like corals with lithified sediment between is much higher than in the rocks from the wall. In addition several specimens of *Acropora palmata* as well as numerous large sticks of *A. cervicornis* were

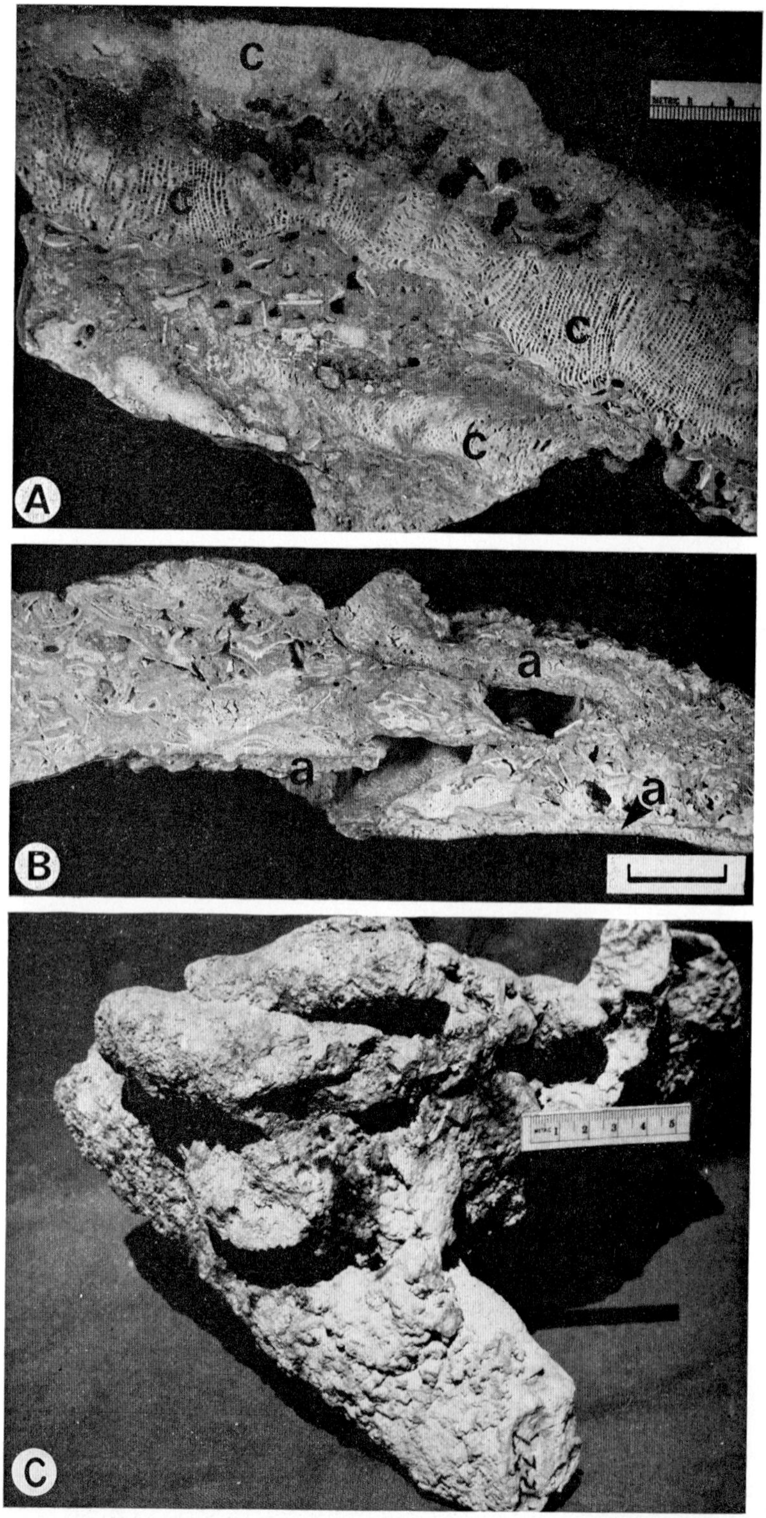
c
c
c
c
A
a
a
a
B
1 2 3 4 5
C

collected. *A. palmata* does not occur at any other localities, and *A. cervicornis* is only present as small sticks elsewhere. Like other limestones, these lithified sediments and corals are altered by numerous generations of cavity formation, sediment infill, and lithification. Many of the cavities at this depth are lined with iron–manganese coatings.

Radiocarbon ages of the limestone

Shallow, poorly cemented sediment above the top of the wall has a radiocarbon age of *c* 4500 years B.P. The bulk of the wall limestone has ages between 7900 and 15 000 years B.P.: two specimens from the surface of the wall have younger ages, *c* 2500 years B.P. and two specimens from the base of the wall have apparent ages of 23 000 and 32 000 years B.P. The age of a specimen from a talus block, and that from the limestone of the furrowed cliffed fore-reef at 174 m, both fall in the same age range as the wall limestones.

Fig. 5-12. Site 7, depth 174 m: (A) a polished slab of limestone from the fore-reef zone composed of three corals [C], all plate-like and between which there is *Halimeda*-rich sediment. The sediment between the upper two corals is extensively bored; scale in cm. (B) A polished slab of limestone from the fore-reef zone composed of three plate-like colonies of *Agaricia* spp. [a] all of which are extensively bored. The sediment sandwiched between the corals is multi-generation *Halimeda*-rich grainstone to packstone; scale bar 2·0 cm. (C) A sample recovered from the blast site and composed of several 'logs' and branches of *Acropora palmata* cemented together and covered with mudstone; scale in cm.

Chapter 6

Petrography of limestones from the wall and fore-reef

INTRODUCTION

In the preceding chapter we have described the composition of limestones from the lower part of the reef front, wall and fore-reef which serves as background for a description of the more intricate, petrographic characteristics of these rocks. In this chapter we focus on the mineralogy, shape and occurrence of the various cements together with the composition of lithified sediments.

CARBONATE CEMENTS

Aragonite and Mg-calcite are the only two cements in these limestones, with Mg-calcite twice as abundant as aragonite. These cements, precipitated between grains of skeletal sand and silt or growing into cavities, display a variety of forms, ranging from micrite-size crystallites to blades and needles of spar. Mg-calcite cement is present as isopachous rinds, either around grains or on cavity walls, and different cement morphologies are produced by variations in the size, geometry and spatial relationships of crystals in these rinds. Aragonite, in contrast, never forms isopachous rinds, but

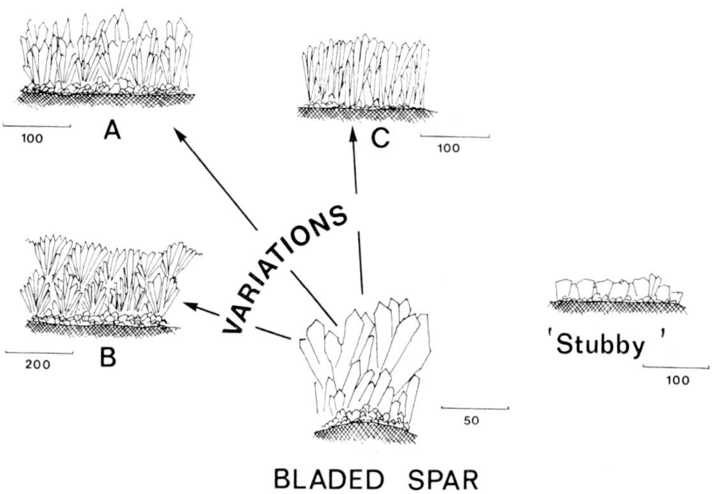

Fig. 6-1. Sketches of the different types of Mg-calcite bladed spar cement; (A) small splays nucleated at separate sites along the grain boundary; (B) two layers of splays; (C) very closely nucleated splays resulting in growth of very elongate blades. The 'stubby' spar at right generally develops in small pores. Scale in μm.

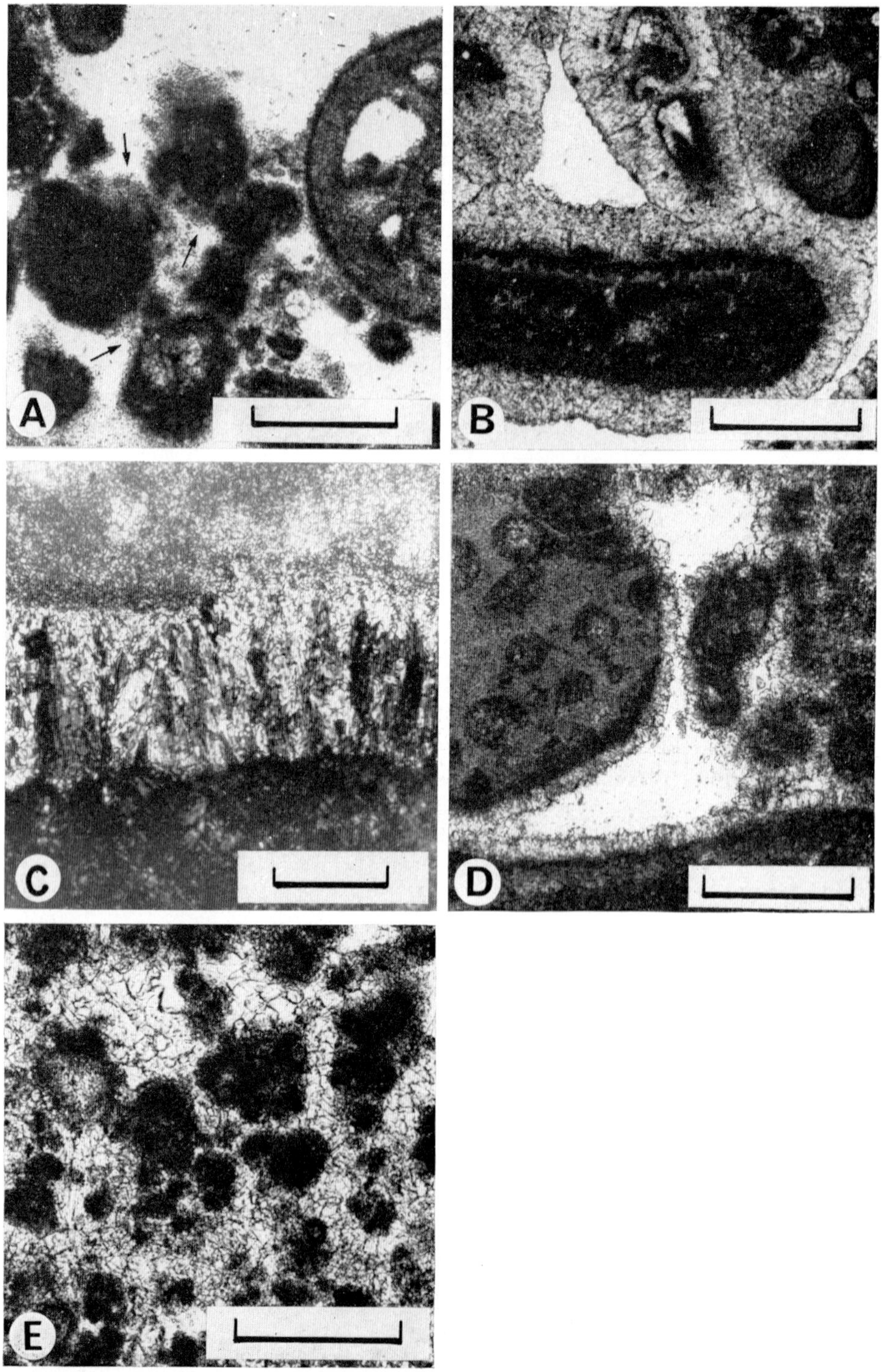
A
B
C
D
E

instead the crystals either extend out from grain surfaces in all directions producing a mesh-like intergranular cement or occur as spherulites expanding rapidly into open cavities.

Mg-calcite cement

Grains are cemented together by, and cavities rimmed with, two styles of Mg-calcite cement, micrite and bladed spar (Fig. 6-1). These two cements are quite distinct; in micrite the equant crystallites are less than 3 μm in size; in spar the rhombic, elongate crystals of clear to cloudy calcite range in length from 40 to 100 μm. Mg-calcite micrite is twice as abundant as bladed spar. These two basic cements that have been described from most subsea cemented limestones (Ginsburg, Shinn & Schroeder, 1968; Ginsburg, Schroeder & Shinn, 1971; Land & Goreau, 1970; Friedman, Schneidermann & Gervitz, 1974) are the two varieties of Mg-calcite found in shallow Belize reef-limestones (James *et al.*, 1976).

Mg-calcite micrite

Small, slightly rounded rhombs of micrite-size Mg-calcite are the most common of all cements (Fig. 6-1A). The cement is precipitated on grain surfaces as isolated crystals that, with time and continued precipitation, form a cement rind of crystallites piled one on top of the other that may be as thick as 25 μm. Micrite cement is ubiquitous between grains of silt, common between grains of mixed sand and silt but rare in sorted sands. In most sediments micrite is the first cement to be precipitated and is followed by all of the other styles, including aragonite.

Mg-calcite spar

All of the Mg-calcite spar in these rocks is composed of one crystal morphology, a steep-sided rhomb that grows out perpendicular to the grain surface or cavity wall and may or may not have pyramidal terminations. This crystal form, recognized as bladed spar from shallow Belize limestone (James *et al.*, 1976), is shaped much like a Roman short sword, gradually expanding from the origin along its length to a maximum and then ending quickly as an obtuse pyramid. The crystal thickness, crystal spacing and pore size of the sediment, however, lead to a variety of different crystal arrangements and therefore different cement morphologies (Fig. 6-1B to E).

Intergranular pores between sand grains, walls of eroded cavities and walls of skeletal pores are often lined with isopachous rinds of bladed spar (Fig. 6-2B).

Fig. 6-2. Mg-calcite cements, site 1, depth 110 m: (A) Photomicrograph in plane-polarized light of friable, medium-grained sand cemented by Mg-calcite micrite; the early stages in the development of micrite rinds are seen in the triangular pore at left (arrow); scale bar 500 μm. (B) Photomicrograph in plane-polarized light of a grainstone composed of *Halimeda* plates, coralline algae fragments and benthic forams cemented by isopachous rinds of bladed Mg-calcite spar; scale bar 500 μm. (C) Photomicrograph in cross-polarized light of one side of an interparticle pore in a medium-grained skeletal grainstone with a cement rind of Mg-calcite composed of bundles of elongate crystals; scale bar 100 μm. (D) Photomicrograph in plane-polarized light of a medium to coarse-grained grainstone cemented by rinds of 'stubby' Mg-calcite spar; scale bar 500 μm. (E) Photomicrograph in plane-polarized light of a small cavity filled with mudstone composed of silt-sized peloids of Mg-calcite, cemented by blocky calcite spar to give a cement similar to the blocky spar of ancient grainstones; scale bar 500 μm.

Individual rinds range in thickness from 20 to 300 μm and show a clearly bladed to diffuse fibrous fabric under crossed nicols (Fig. 6-2C). Individual crystallites vary in length from 40 to 100 μm but are remarkably consistent in terminal width, varying between 5 and 7 μm, no matter what the locality of the sample.

The well developed blades of Mg-calcite are best seen when the cement is developed around a curved substrate and the cement fringe is a radial array of blades with the whole structure exhibiting sweeping extinction.

Isopachous rinds are composed, not of blades growing perpendicular to the substrate, but of a series of partial spherulites or splays, nucleated at point centres 60–150 μm apart along the substrate. Crystals at the periphery of each splay interfere with crystals growing from adjacent splays so that competitive growth favours development of those crystals in the central portion of the splay (Fig. 6-2C). Consequently, the cement rind is composed of many adjacent splays each with a sweep of from 30 to 90°. Each individual splay has sweeping extinction, and where crystals can be recognized, each crystallite also has sweeping extinction.

In most instances individual crystallites are extremely difficult to differentiate in the cement rinds, because they are so small and are in optical continuity with their neighbours. When cement rinds are thick and centres of nucleation close together, so that many crystals are almost parallel to the substrate and crystals are in good optical continuity, the overall rind has the appearance of fibrous calcite, composed of many acicular submicroscopic crystallites.

The commonest arrangement (Fig. 6-1) is that of splays, nucleated at, or close to, the grain boundary. In some thick rinds individual generations are separated by thin zones of micrite cement, and a new series of splays grows outward from the micrite. Alternatively, there may be more than one generation of splays without discontinuity between them and in these the second generation is nucleated between the splays of the first generation, giving the rind a complex sweeping extinction and enhancing the fibrous appearance.

An uncommon, but persistent arrangement is that of slightly thicker crystals, up to 10 μm wide, within closely spaced splays so that the majority of the crystallites grow perpendicular to the substrate. These crystals can be easily differentiated as each has individual extinction and the whole rind has a twinkling as opposed to fibrous appearance.

The gradually widening, bladed nature of the crystals has been described by Schroeder (1972) on calcified algal filaments in Bermuda Cup reefs, where the crystals' morphology appears identical to the cement described here, and from shallow water Belize reefs (James *et al.*, 1976). Schroeder (1972), however, does not emphasize the spherulitic nature of the cement in other cement fringes but instead observes a 'palisade' cement, stressing the parallel nature of the crystals. Examination of the Belize limestones suggests that the palisade cement is spherulitic.

One conspicuous variant of bladed spar is a small 'stubby' crystal that forms isopachous rinds of cement, particularly around silt-sized grains and in interparticle pores less than 50 μm wide (Fig. 6-2D). The length to width ratio of these crystals is relatively small, ranging from 1 : 1 to 2 : 1. The width of the crystals is almost as large as the widest blades in isopachous rinds, 8 μm, but the crystals are rarely longer than 20 μm. When facing open voids such crystals widen away from the substrate and terminate either in the familiar obtuse pyramid or flat pinacoid. In silt-size sediment, or fine sand-size sediment, where the pore sizes are generally between 20 and 100 μm, the cements do not have room to develop fully and so grow out from grain surfaces to meet similar 'stubby' spar growing from the opposite grain (Fig. 6-2E).

Close examination of what appears to be rhombic crystals in some voids reveals that they are crystals of 'stubby' bladed spar. In some rare pores, however, unequivocal blocky, equant spar is present. This cement form occurs as a mosaic of irregular rhombs with no preferred orientation to the substrate. Individual rhombs range from 4 to 10 μm in size and average 6 to 8 μm; they either occur alone or as the last stage of cement adjacent to a large, open void.

Aragonite cement

Aragonite cement characteristically forms long needle-like crystals identical to aragonite cements found in synsedimentary limestones from many shallow sedimentary environments (Ball, 1967; Land & Goreau, 1970; Shinn, 1969; Ginsburg *et al.*, 1971; Friedman, Amiel & Schneidemann, 1974), as well as the shallow marginal reefs of Belize (James *et al.*, 1976).

In limestones from wall, these needle-like crystals are precipitated in a variety of forms, as a mesh of crystals between particles, as epitaxial overgrowths on particular aragonite skeletons, as botryoidal splays in cavities of all kinds and rarely as equant blocky crystals (Fig. 6-3).

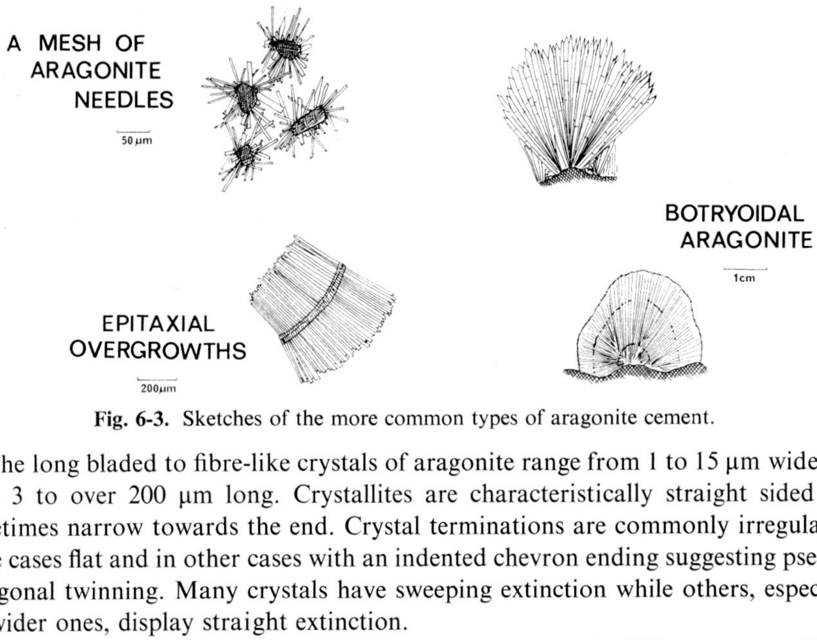

Fig. 6-3. Sketches of the more common types of aragonite cement.

The long bladed to fibre-like crystals of aragonite range from 1 to 15 μm wide and from 3 to over 200 μm long. Crystallites are characteristically straight sided and sometimes narrow towards the end. Crystal terminations are commonly irregular, in some cases flat and in other cases with an indented chevron ending suggesting pseudo-hexagonal twinning. Many crystals have sweeping extinction while others, especially the wider ones, display straight extinction.

Aragonite occurs as both an intergranular and cavity-filling cement. When intergranular, the aragonite occasionally succeeds a thin rind of Mg-calcite micrite. On the whole, cemented sediments contain an average of 10–20% aragonite cement, but in some *Halimeda* grainstones the cement may be as much as 50% aragonite.

A mesh of aragonite needles

The commonest form of intergranular aragonite cement is an irregular chaotic arrangement of needle-like crystals forming a mesh of aragonite needles growing out from grain surfaces (Fig. 6-4A). These crystals grow in two ways, either as a splay or

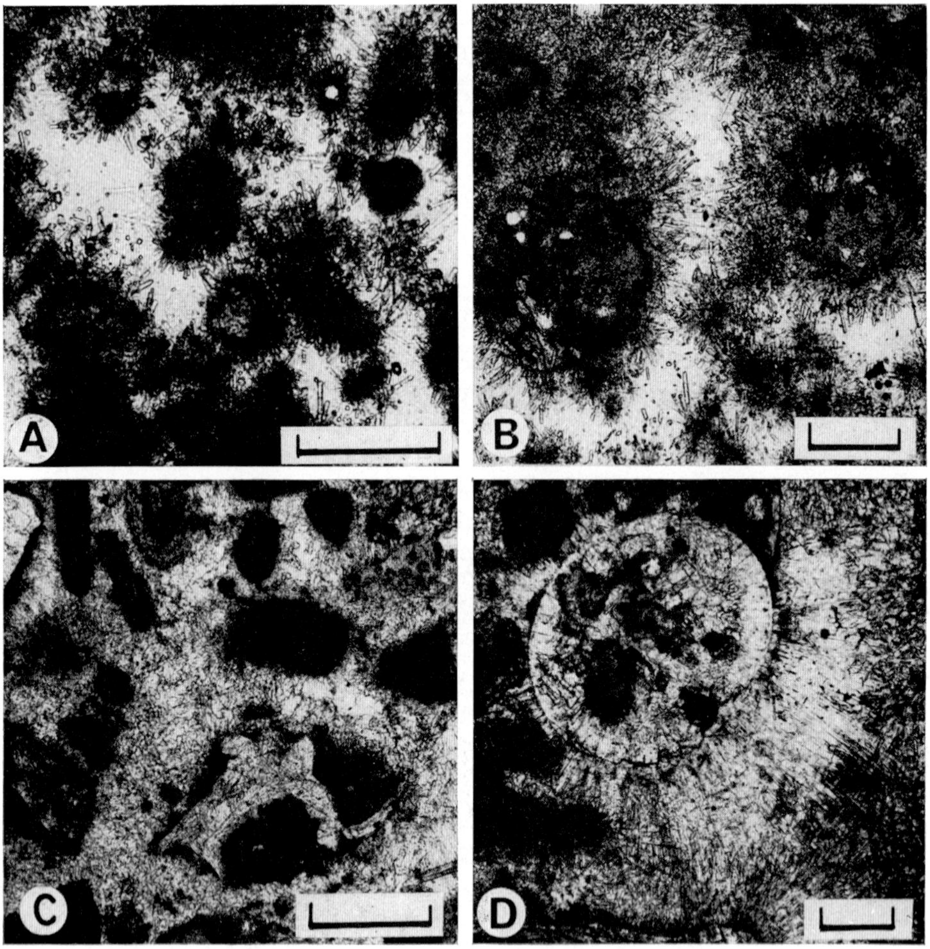

Fig. 6-4. Aragonite cements. Site 1, (A) depth 110 m: photomicrograph in plane-polarized light of silt in which the widely spaced grains are poorly cemented by aragonite in the form of needles of various shapes growing outward from grain surfaces in all directions; scale bar 500 μm. (B) Depth 100 m: photomicrograph in plane-polarized light of sediment composed of sand and silt-sized skeletal grains cemented by aragonite needles that interfere in interparticle pores to form a mesh of needles; scale bar 200 μm. (C) Depth 110 m: photomicrograph in plane-polarized light of a grainstone well cemented by a mesh of aragonite needles; scale bar 500 μm. (D) Depth 88 m: photomicrograph in plane-polarized light of a skeletal grainstone in which aragonite cement has grown on a gastropod shell, with crystals of cement in optical continuity with the crystals that form the shell wall; scale bar 200 μm.

fan of many crystals radiating out from a single point or as individual crystals with no preferred orientation. The most abundant form, random individual needles, grows irregularly with adjacent needles rarely the same width or length, often eight to ten times different in size, and commonly in directions that may be as much as 90–120° apart (Fig. 6-4A, B). In most instances the individual, randomly oriented crystallites appear to grow most rapidly and interfere with adjacent spherulite growth so that the eventual result is a mesh of overlapping and interlocking crystals filling the pore (Fig. 6-4B, C). There are commonly minute pore spaces throughout the cement where fibres of aragonite have crossed, but the spaces between fibres are not filled. Individual crystallites in this cement mesh are from 1 to 6 μm wide and of varying length, depending on the size of the void, but commonly between 15 and 45 μm.

Epitaxial cement overgrowths

In contrast to the mesh of crystals between most particles, relatively large sweeping arrays of crystals sometimes extend out in optical continuity from specific aragonite particles. The cement forms most commonly on small gastropods, as well as on bivalve fragments, where the crystals of the host are oriented normal to the skeletal wall (Fig. 6-4D). Aragonite grows in optical continuity with the skeleton into surrounding voids and resembles fibrous aragonite spherulites described below because it is formed of many very small crystallites. This cement is analogous to calcite cement precipitated on echinoid plates and spines found in ancient carbonates. Aragonite crystal growth is clearly enhanced in this situation as such epitaxial cements are by far the best developed in otherwise poorly cemented sediments. In firmly cemented calcarenites, epitaxial cement on small gastropod fragments always covers an area far in excess of the volume of gastropod grains. These epitaxial cements are commonly surrounded by aragonite mesh cement.

Botryoidal aragonite

The largest and most spectacular cements are spherulites or fans of aragonite which we have called botryoidal aragonite and described in detail elsewhere (Ginsburg & James, 1976). These growths develop either when splays growing as interparticle cement expand into an adjacent cavity (Fig. 6-6B) or when splays nucleated at specific points on cavity walls are the only cements growing in the cavities (Fig. 6-5A, B).

In many specimens from the wall there are botryoidal growths lining and filling voids. These crystal arrays so closely resemble some cave deposits that, when first discovered, they produced a major controversy aboard ship. The cement is often seen easily in hand specimen and grows from the walls of cavities as single to multiple and often superimposed spherulites. Individual structures of botryoidal aragonite range from less than a millimetre in radius to more than 5 cm (Fig. 6-5A); the largest array of crystals fills a cavity over 25 cm across. Individual and coalescing botryoidal aragonite occurs in all kinds of cavities: original coral pores, small borings of *Cliona* sponges, cylindrical bivalve borings (Fig. 6-6A), large irregular excavations of *Siphonodictyon* sponges (Fig. 6-5), shelter cavities formed by round and platey corals and irregular cavities of unknown origin (Fig. 6-7A, B). The orientation of the cement is not geostrophic but instead it grows into a cavity from any direction (Fig. 6-5B). Precipitation is clearly synchronous with sedimentation as geopetal sediments and cement are interlaminated on cavity floors. Precipitation is not restricted to a single episode; in several samples the cement has been bored by endolithic organisms and the holes filled with sediment, which is, in turn, overlain by more botryoidal aragonite.

In thin section it can be seen that botryoidal aragonite is not a unique cement but is one isolated, albeit spectacular, growth of common aragonite spherulites that occurs in very small pores and between particles. The crystal fabric of the botryoidal growths and spherulites ranges from innumerable long but minute crystallites, less than a micron wide in structures that have sweeping extinction (Fig. 6-6D), to many

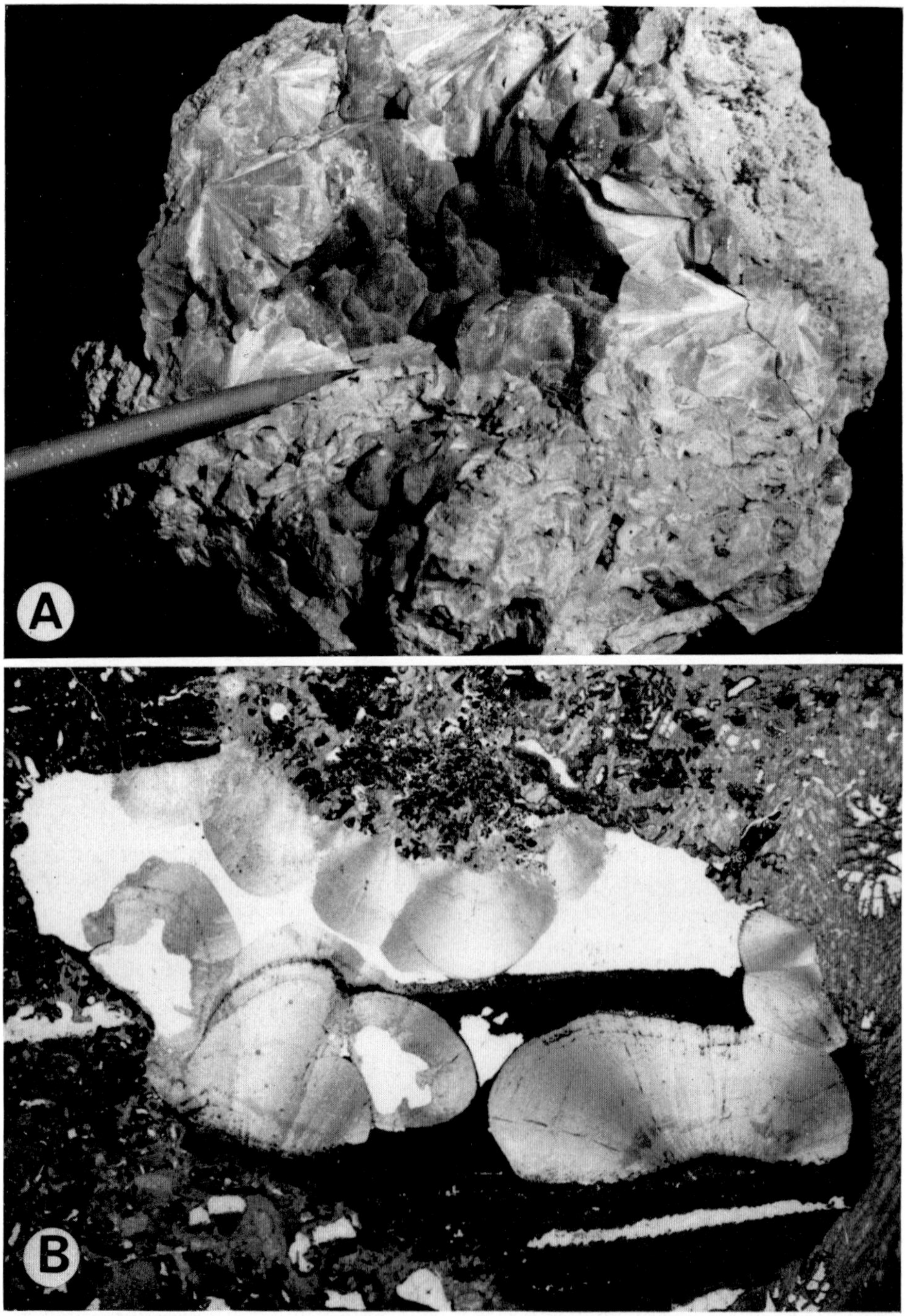

Fig. 6-5. Botryoidal aragonite: Site 1, (A) depth 143 m: a limestone composed of coral and mudstone and containing a large cavity much of which is filled with growths of botryoidal aragonite; the pencil points to geopetal marine sediment overlying the cement at base. (B) Depth 110 m: photomicrograph in partially polarized light of a cavity eroded into a limestone composed of coral (upper right) and mudstone (lower left) in which botryoidal aragonite has grown, both from the ceiling and floor of the cavity, as well as at different stages when sediment was filling the void; note the layers of sediment in the crystal arrays indicating that sediments were filtering in while the cement grew (holes in the sediment and aragonite are artifacts of thin section preparation); scale bar 1·0 cm.

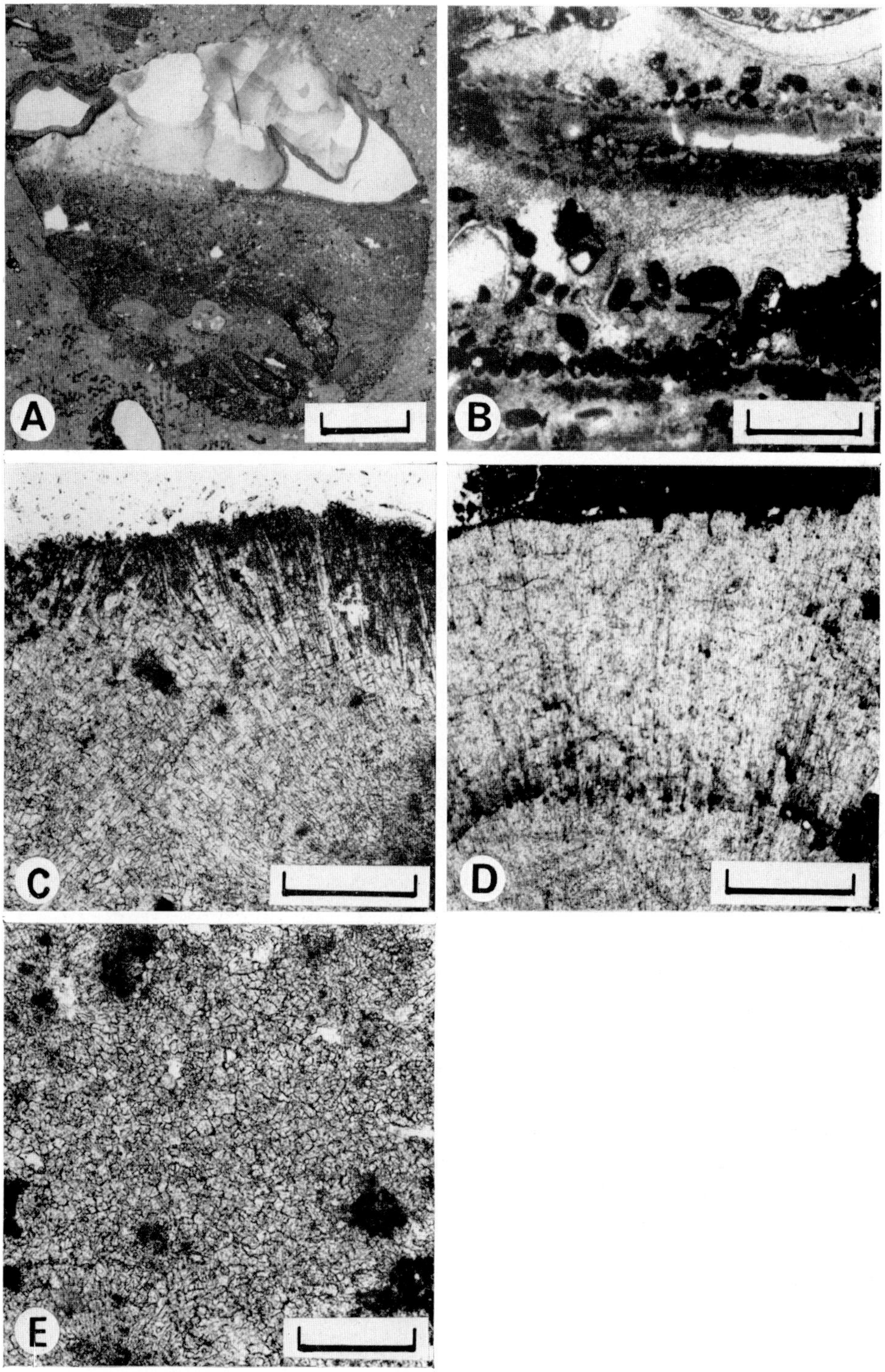
A
B
C
D
E

long, clearly defined crystals each of which is from 2 to 15 μm wide and may be up to 200 μm long in structures that have twinkling extinction (Fig. 6-6C). Botryoidal arrays formed of the larger crystallites never seem to exceed 0·5 mm in radius whereas those with minute crystallites may be 10 cm in radius.

The amount of botryoidal aragonite was enough for ^{14}C age determinations (Table 6-1). Two independent ^{14}C age determinations on one sample (Fig. 6-5) yielded 12 740 ± 250 years B.P. (G. Stipp, University of Miami) and 12 600 ± 150 years B.P. (G. Otslund, University of Miami) within the age range of 7000–15 000 years B.P. of coral and sediment samples from the wall. The carbon and oxygen isotope values of the aragonite, when plotted on a diagram in which 480 isotope ratios of modern and Pleistocene carbonates are arranged into different fields, fall in the small area occupied by values obtained from other subsea cements (Fig. 6-8). The strontium content of the aragonite is the same as that in aragonite precipitated, both inorganically and organically (except for molluscs), from sea water (7700–10 000 p.p.m.), and different from the amount of strontium in aragonites precipitated from fresh waters (*ca* > 1000 p.p.m.) (Kinsman, 1969) (Fig. 6-9A). If the Sr in the aragonite is used as a geothermometer an average value of 8285 p.p.m. would indicate precipitation from sea water around 26·5°C; ambient temperatures between 70 and 140 m are today somewhat cooler at 21–25°C. The sodium content of the aragonites is well above 1000 p.p.m. (Fig. 6-9B), the dividing line between those aragonites precipitated from fresh water (less than 1000 p.p.m.) and sea water (more than 1000 p.p.m.) (Land & Hoops, 1973).

Taken individually and collectively all lines of evidence, petrographic, isotopic age, isotopic, and trace elements show that the botryoidal aragonite was precipitated from sea water and there was no dilution by percolating fresh groundwaters.

Blocky aragonite crystals (Fig. 6-6E)

A volumetrically unimportant, but unusual form of cement is a series of interlocking, equant to irregular crystals that forms in intergranular voids between the other aragonite and Mg-calcite bladed cements. These crystals become very common near the top of geopetal fills and form a granular cement. The individual crystals are 3–15 μm across and have a length to width ratio of 1 : 1–3 : 1. The crystal boundaries are often sharp, but are occasionally embayed and irregular; individual crystals have sweeping extinction, and they grade in any direction towards the more elongate aragonite mesh cement.

Fig. 6-6. Aragonite cements. Site 1, depth 110 m: (A) photomicrograph in partially polarized light of a colony of *Porites astreoides* that has been bored by the date mussel *Lithophaga* sp. forming the elliptical hole at centre. The boring is partially filled with sand and mud, cemented both by Mg-calcite micrite (dark areas) and by aragonite (light areas) while botryoidal aragonite grows from the roof and sediment surface and is succeeded by a rind of Mg-calcite bladed spar; scale bar 0·5 cm. (B) Photomicrograph in plane-polarized light of *Halimeda* grainstone in which particles are cemented by a mesh of aragonite needles (left). The aragonite expands into a cavity at centre and changes to growths of spherulitic aragonite. This last stage is overlain by Mg-calcite cement, right centre; scale bar 500 μm. (C) Photomicrograph in plane-polarized light of a large intergranular pore into which two aragonite spherulites have grown and interfered. The upper part of this cement is illustrated here, composed of botryoidal aragonite comprising easily identifiable crystals succeeded by a later stage of Mg-calcite micrite cement (top); scale bar 500 μm. (D) Photomicrograph in plane-polarized light of an aragonite botryoid composed of innumerable small needles with a layer of Mg-calcite peloids and small skeletal particles in the middle of the structure; scale bar 500 μm. (E) Photomicrograph in plane-polarized light of a rare example of a mosaic of blocky aragonite cement and particles of silt; scale bar 200 μm.

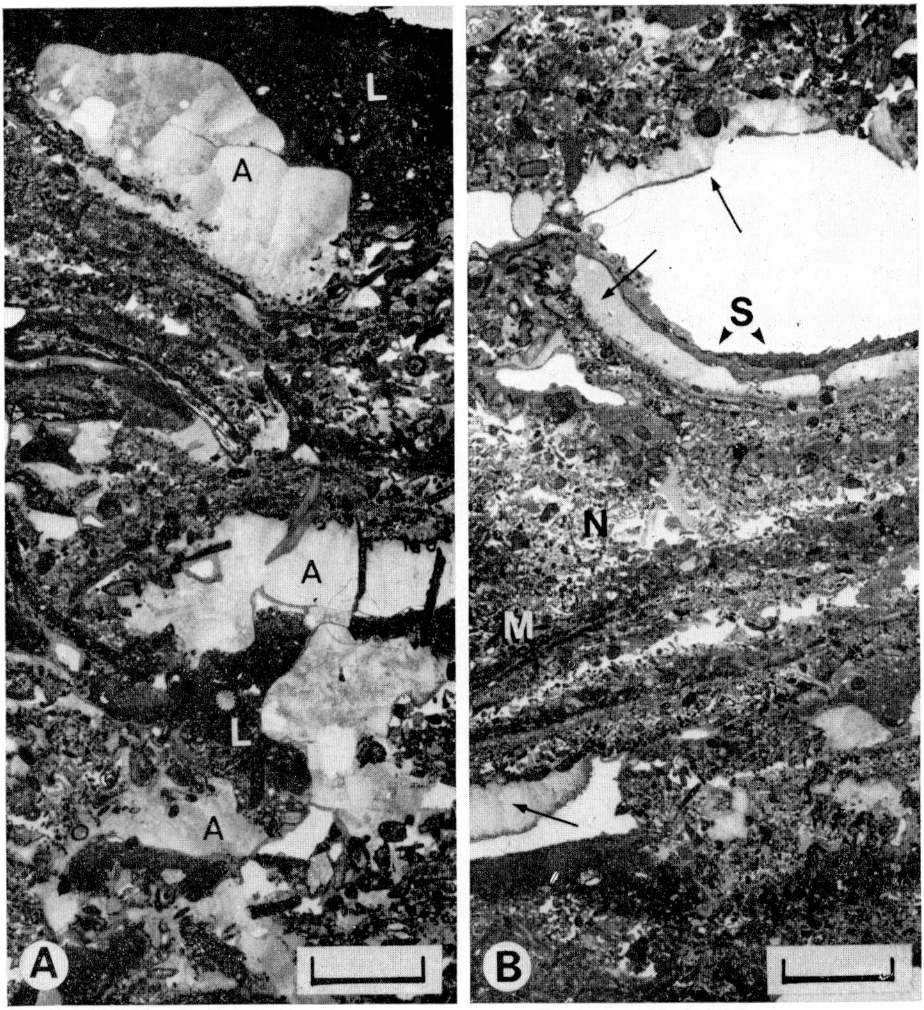

Fig. 6-7. Botryoidal aragonite cements in grainstones, site 1, (A) depth 110 m: a photomicrograph in partially polarized light of a *Halimeda*-rich grainstone to packstone in which botryoidal aragonite [A] has grown into voids. The cement is commonly overlain by silt [L] and succeeded by a thin rind of Mg-calcite cement (dark rind); scale bar 0·5 cm. (B) Depth 88 m: photomicrograph in partially polarized light of grainstone to packstone in which some areas are not cemented [N], some areas are cemented by Mg-calcite micrite cement [M], and some cavities are partially filled with growths of botryoidal aragonite (arrows); botryoidal aragonite is often overlain by sediment [S]; scale bar 0·5 cm.

Table 6-1. Geochemical analyses of botryoidal aragonite

Sample No.	$\delta^{13}C$	$\delta^{18}O$	Sr^{2+}	Na^{+}
	(‰ vs PDB)			
72.23.9	+3.3	+0·6	8120±25	3100
72.23.14	+4·3	+0·4	8460±20	2700
72.23	+4·2	+1·0	—	—
72.24.48	+4·2	+1·0	8390±90	3600
72.29.2	+4·2	+1·1	8170±155	3200

Analyses supervised by P. W. Choquette, Marathon Oil Company; Sr^{2+} determinations by atomic absorption: Na^{+} by emission spectrometry.

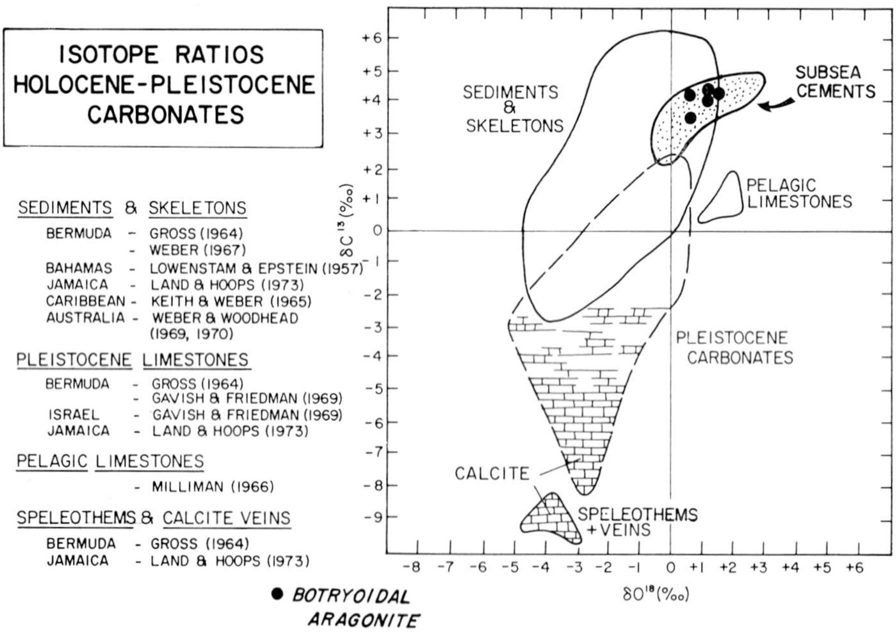

Fig. 6-8. The carbon and oxygen isotope ratios of various skeletons, sediments and cemented limestone of Holocene and Pleistocene age (the sources of the 463 data points used to outline the areas are tabulated at left). The data to delimit the area of subsea cements comes from studies in Bermuda (Ginsburg, Schroeder & Shinn, 1971), Jamaica (Land & Goreau, 1970) and the Persian Gulf (Shinn, 1969); Ratios versus the PDB standard.

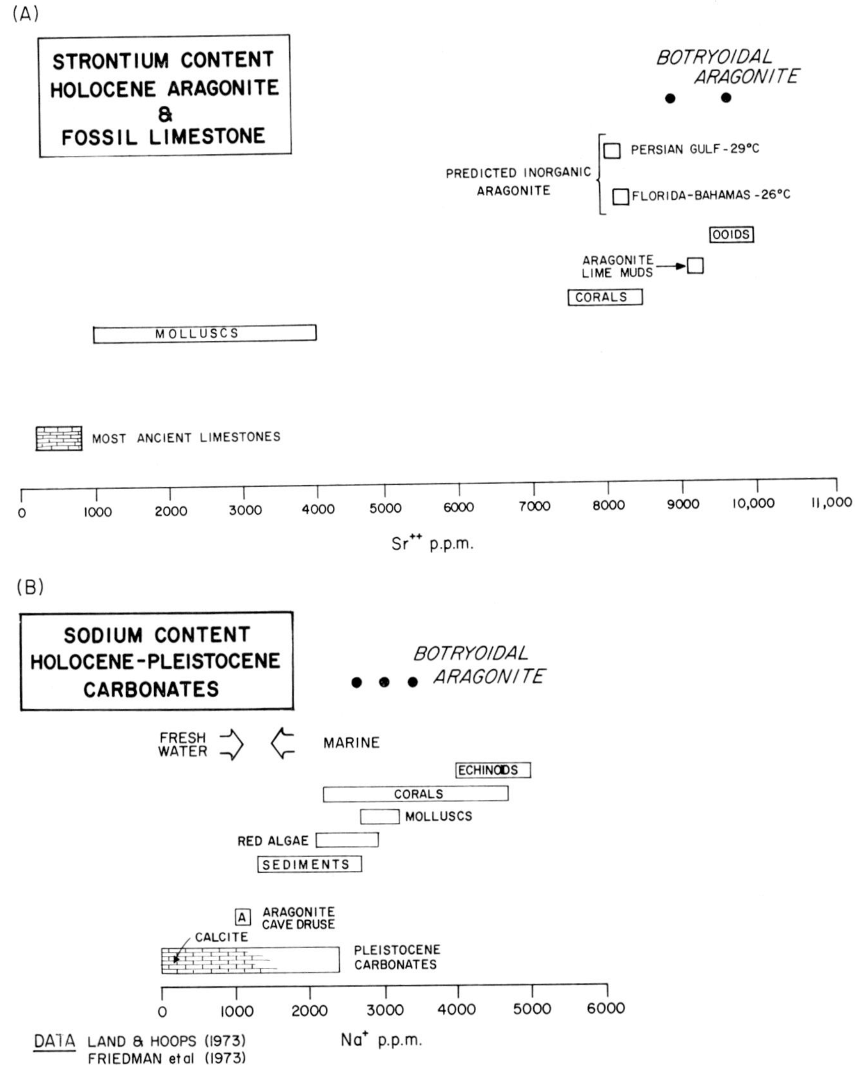

Fig. 6-9. (A) A diagram illustrating the strontium (Sr) content of aragonite precipitated both inorganically and as skeletons in Holocene marine environments as well the strontium in Pleistocene limestones that have altered to calcite (data from Kinsman, 1969). (B) A diagram illustrating the sodium (Na) content of carbonate sediments and skeletons of Holocene and Pleistocene age (data from Land & Hoops, 1973; Ameil *et al.*, 1973).

HOLOCENE AND LATE PLEISTOCENE LIMESTONES

Halimeda grainstones to wackestones

Composition

Halimeda-rich limestones range in texture from homogenous *Halimeda* calcarenites and calcilutites to irregular 2–3 cm thick laminations of calcarenite and calcilutite to burrowed and bioturbated sediment. These sediments span the range in

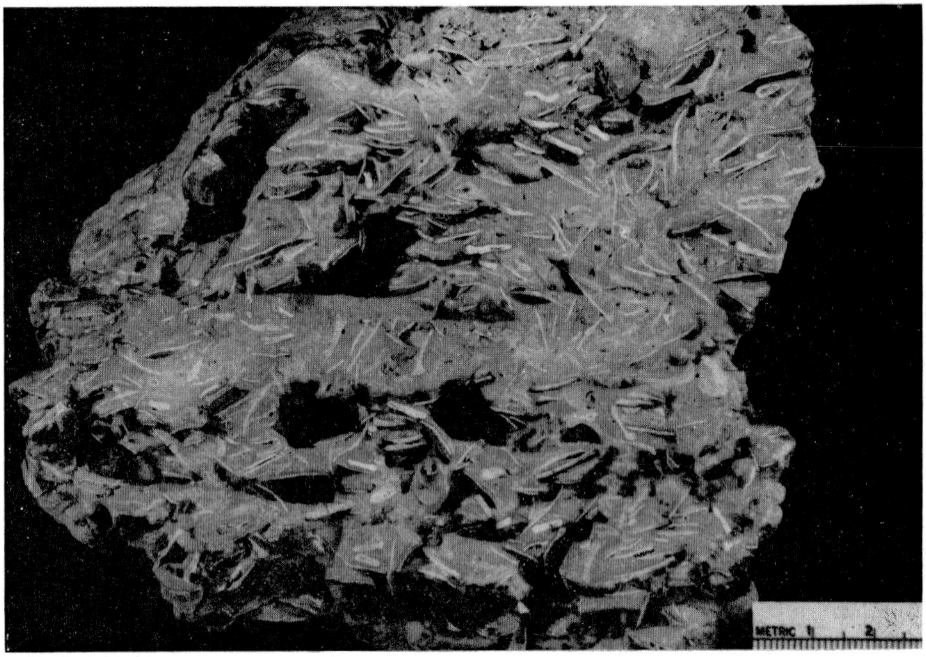

Fig. 6-10. Site 1, depth 110 m: a polished slab of *Halimeda* wackestone with many *Halimeda* plates forming numerous shelter voids, many of which are lined with red-brown iron-oxide coatings; scale in cm.

texture from grainstones to packstone to wackestones (Dunham, 1962). The arrangement of *Halimeda* plates within these sediments ranges from a self-supporting 'house-of-cards' structure (Fig. 6-10) with finer sediments partially or wholly filling the spaces between plates to a thorough mixture of *Halimeda* plates and other sediments to *Halimeda* plates floating in a matrix of fine sand and mud-sized carbonate (Fig. 6-11).

The large, flat *Halimeda* plates often form such an interlocking structure that it is clear they were deposited before the fine sediment between them. That this fine-grained sediment is internal is shown by its geopetal fabric, often with good shelter porosity still present (Figs 6-11B, 6-12). The internal sediment of partially filled spaces between plates is most commonly fine to medium-grained sand (Figs 6-11C, D, 6-12). If the spaces are almost completely filled, the sediment is most commonly sand at the base and mud at the top, sometimes graded (Figs 6-12, 6-13D) and in other instances with a clear discontinuity between sand and mud. The mud that has 'topped-off' a cavity filling is often up to half of the sediment in any one cavity. Even in sediments that are a mixture of *Halimeda* plates and sand, where there is little evidence of cavity filling, the small cavities beneath plates are partially to completely filled with geopetal mud.

Although the fabric of *Halimeda* plates with geopetal sediment between indicates internal sedimentation, the fabric of mixtures of *Halimeda* plates and sand or mud may be produced by bioturbation. Some samples are subtly mottled with disrupted laminae and others contain clots of mud-rich sediment. Other specimens show large

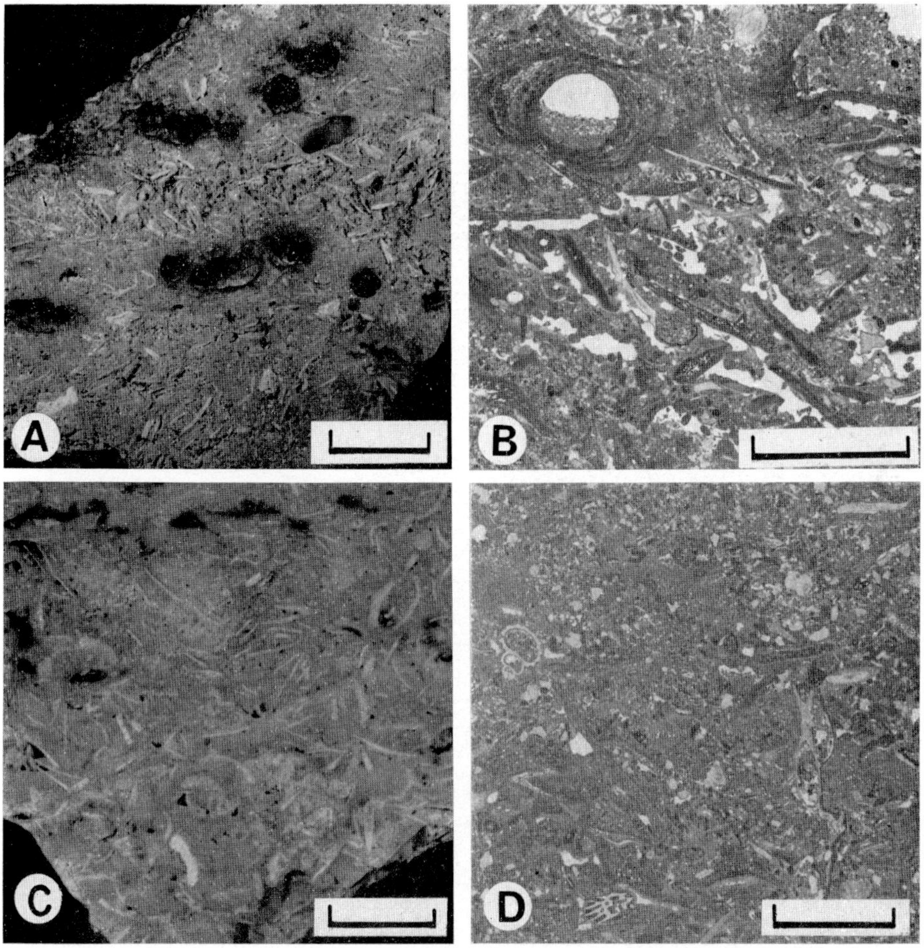

Fig. 6-11. *Halimeda*-rich sediments, Site 1, depth 110 m: (A) a polished slab of well burrowed *Halimeda* grainstone to packstone, note that burrows appear to be concentrated in the irregular layers with fewer *Halimeda* plates and more mud; scale bar 2 cm. (B) A photomicrograph in plane-polarized light of *Halimeda* grainstone with several burrows, one of which is illustrated here; each burrow has several concentric coatings or linings of silt and the vacated cavity is now floored with a geopetal filling of silt. The *Halimeda* plates act to form numerous small shelter cavities; scale bar 1·0 cm. (C) A polished slab of *Halimeda*-rich packstone to wackestone with almost all of the shelter voids filled with silt; scale bar 2·0 cm. (D) a photomicrograph of the sample illustrated in (C) with most of the spaces between sand-sized particles filled with silt; scale bar 0·5 cm.

sinuous open burrows (Fig. 6-11A, B) ranging from 4 to 5 mm in diameter with walls of mud formed of two to seven concentric laminae and they may be empty or partially filled with lithified, internal mud.

The sand between *Halimeda* plates ranges in size from medium to very fine sand grade, with fine-grained sand the commonest. Sand makes up from half to two-thirds of most rocks, but is substantially reduced in calcilutites where it is rarely more than one-third of the sediment. The commonest grains, in about equal proportion, are the small branches of articulated coralline algae, angular coral fragments, pieces of

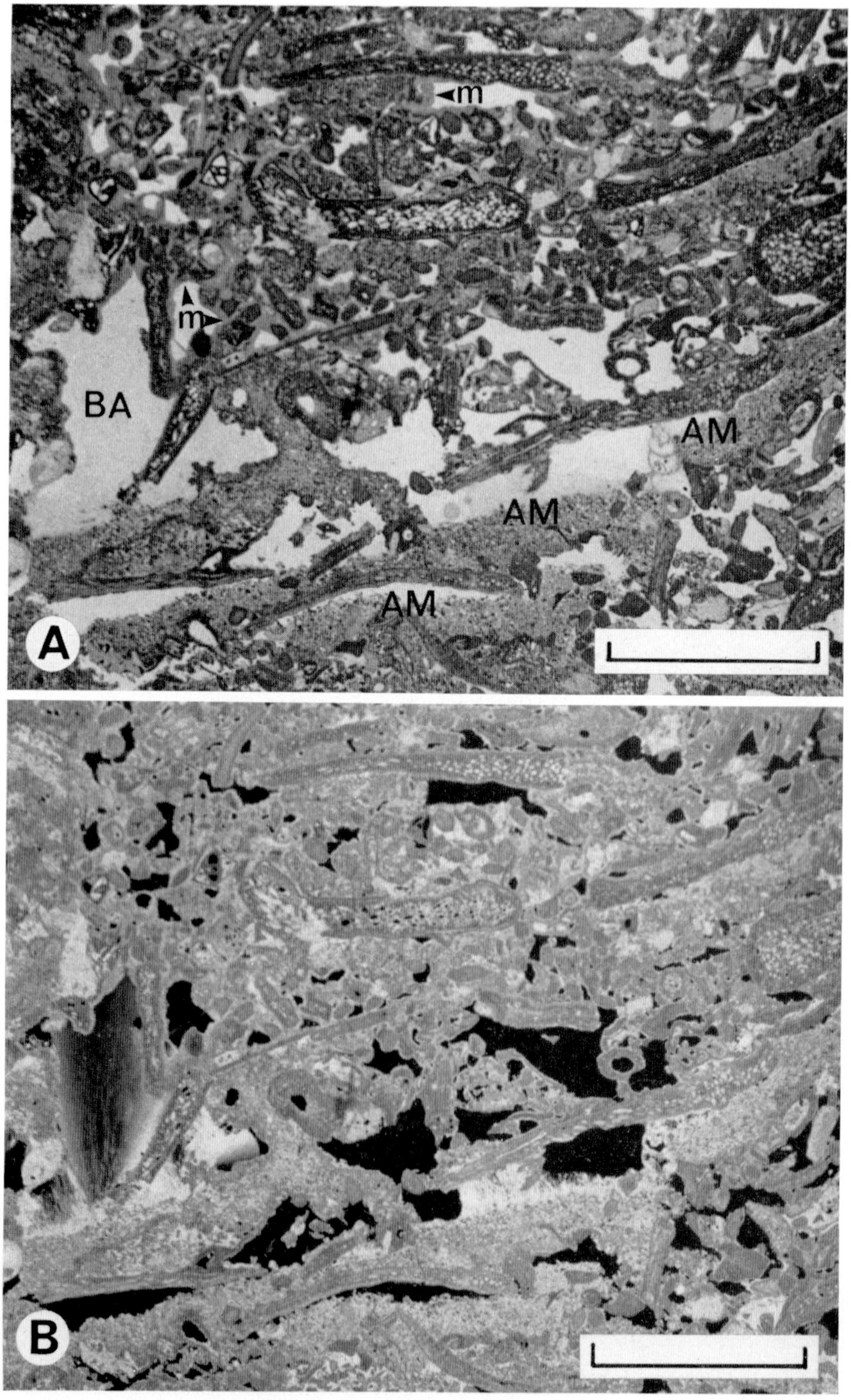
m
m
BA
AM
AM
AM
A
B

mollusc shells, benthonic and planktonic foraminifer tests and rounded to ovoid peloids. Peloids, from 0·3 to 0·5 mm in long dimension, are composed of mud and identical to the material lining the open burrows described above and they alone may make up as much as a third of the sand-sized fraction. Occasional, large, whole *Homotrema* tests, echinoid plates and other larger foraminifera tests can be seen scattered throughout the sediment but are not a major component. In addition, in the very fine sand-sized fraction ostracode carapaces and fragments of crustose coralline algae occasionally are of equal importance to the commonest grains described above.

Mud is about two-thirds silt-size grains and one-third clay-sized particles. Half of the silt is composed of peloids of micrite-sized Mg-calcite, ranging in size from 25 to 70 μm, and of unknown origin (James *et al.*, 1976). The remainder of the silt is composed of irregular aragonite particles and skeletal debris. The aragonite grains are characterized by their consistent size, 30–100 μm, and faceted shapes—a diagnostic form for chips excavated from aragonite substrates by burrowing sponges (Fig. 6-13A) (Warburton, 1958; Reutzler & Rieger, 1973; Futterer, 1974).

Cementation

There is a general relationship between sediment grain size (and therefore pore size) and the mineralogy as well as form of the cements. The common silt-sized grains are almost exclusively cemented either by Mg-calcite micrite (Fig. 6-13A) or short 'stubby' bladed spar and occasionally equant aragonite. Sand-sized grains and coarser particles are cemented by the various aragonite cements (Figs 6-12, 6-13C, D, E, F) as well as bladed Mg-calcite spar (Figs 6-12, 6-13B, E). Because the sediments are so complex both in their depositional and alteration fabrics—mixtures of mud, sand and larger grains as well as skeletons—the cements are equally diverse in any given sample (Fig. 6-12).

In grainstones both cements and grains are easily distinguished, but in silt, because of the fine grain size of the particles and small crystal size of the cements, this differentiation is not always easy and at times impossible in thin section. Much of the silt is micrite peloids (Fig. 6-13B) as well as aragonite chips and other skeletal debris. In friable silts the cements are commonly a thin fringe of Mg-calcite, which is easily seen. If the sediment is well lithified and the grains close together and cemented by Mg-calcite micrite, then cement and grains are difficult to distinguish and the aggregate appears to be a homogenous mud (Fig. 6-13B).

If the pores between silt grains are larger than 15–20 μm, cements of both micrite and stubby bladed spar generally occur as a thin rind of micrite followed by short blades of spar growing out from the grain surface and filling the voids. The last few layers of silt in a cavity filling are well cemented, often by coronas of bladed spar 60–80 μm wide, and the last layer extends up into the void as a continuous rind of cement.

In grainstones and packstones both aragonite and Mg-calcite are important

Fig. 6-12. Site 1, depth 110 m: two photomicrographs, one in plane-polarized light (A) and one in cross-polarized light (B) of a *Halimeda* packstone composed of large *Halimeda* plates and sand-sized skeletal particles cemented by isopachous fringes of Mg-calcite [M] and a mesh of aragonite crystallites [AM] with spherulitic growths of botryoidal aragonite [BA] in some cavities. Note the grading of internal sediment between *Halimeda* plates particularly in the lower half of the photograph; scale bar 0·5 cm.

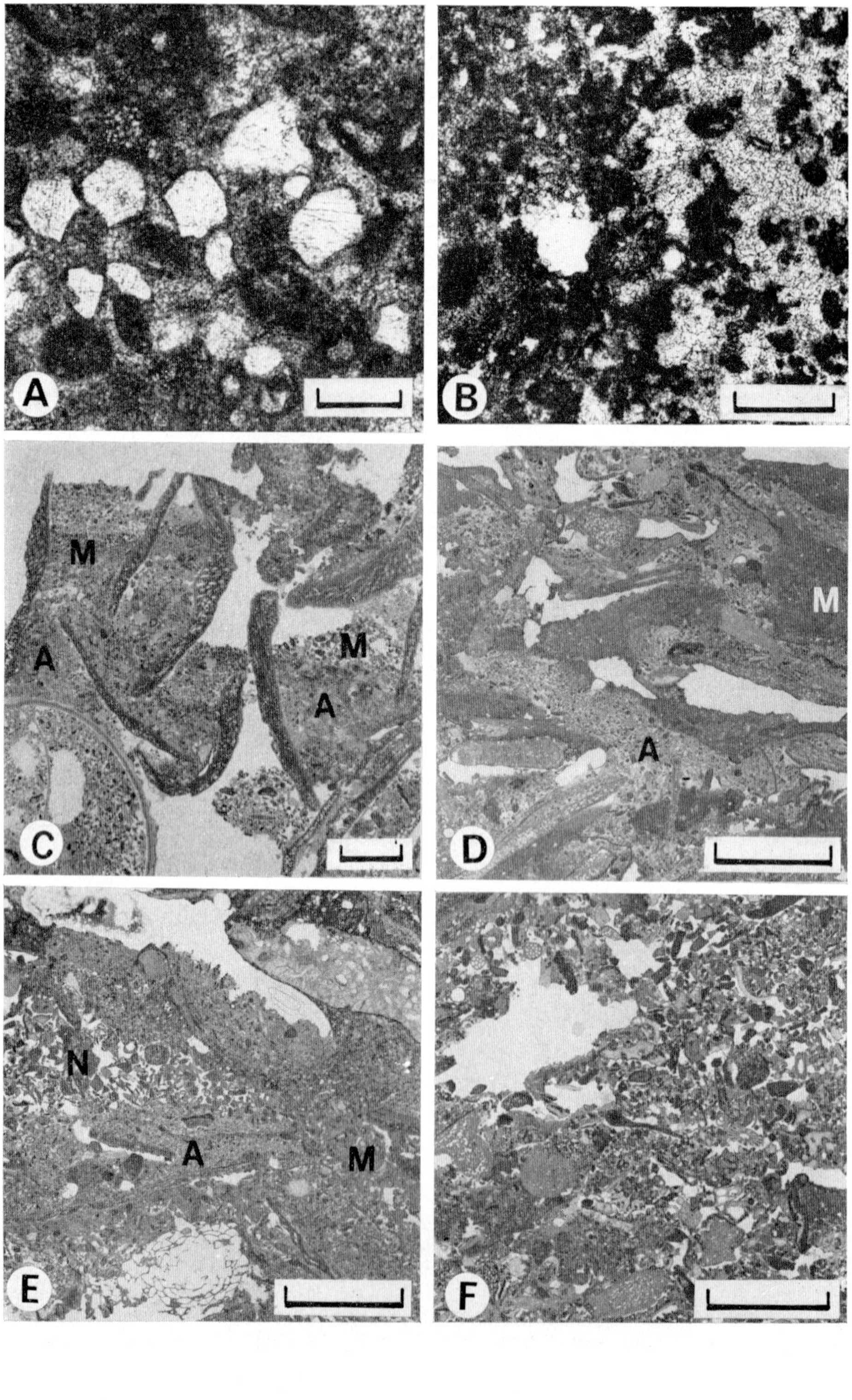
A
B
M
A
M
A
C
M
A
D
N
A
M
E
F

cements. These cements, however, are not always in a sequence but rather present as irregular domains or blotches with one type of cement grading laterally or vertically into another crystal morphology or mineralogy (Fig. 6-13B, C, D). The degree of cementation is also not uniform. Fine-grained sediments, possibly because of the small amount of cement necessary for lithification, are almost always well cemented, while contemporaneous sand-size sediments are often poorly lithified. In addition, borings into poorly cemented grainstones may be lined with several thick fringes of cement (Fig. 6-21A). The degree of cementation may grade from areas of almost no cement to areas with thick intergranular fringes (Fig. 6-13E, F).

Sequence of cementation

Despite the wide depth range, geographical and environmental differences, the *Halimeda*-rich sediments show a consistent sequence of cementation as well as sediment texture from place to place. As outlined above, the *Halimeda* sediments are generally composed of *Halimeda* plates, between which is a partial infill of fine to very fine-grained sand, succeeded by layers of silt. The cementation of these sediments shows a sequence as well. The fine to very fine-grained skeletal sands on top of the *Halimeda* plates are generally cemented by aragonite mesh cement. This relationship is common at all localities, from the shallow, relatively young friable sediments at a depth of 40 m (Fig. 6-13C) to the well cemented *Halimeda* limestones of the wall and ridge and furrow (Fig. 6-13D). The mixed sand and silt, or silt in the upper part of shelter cavities are commonly cemented by Mg-calcite micrite (Fig. 6-13D). Grains and sediment are very often difficult to distinguish, but toward the top of the sediment filling peloids and aragonite sponge chips become more and more distinct because they are cemented more and more by 'stubby' bladed spar. The intervening zone of micrite-cemented silt is not always present and the aragonite cemented grainstones commonly grade up into silt cemented by bladed spar.

Fig. 6-13. (A) Site 1, depth 110 m: photomicrograph in plane-polarized light of silt composed of angular chips of aragonite (light particles) with characteristic scalloped margins that were eroded from a coral or mollusc skeleton by an endolithic sponge. The silt-sized grains are cemented by thin rinds of Mg-calcite micrite that appear dark in the photograph; scale bar 100 μm. (B) Site 1, depth 110 m: a photomicrograph in plane-polarized light of silt composed of peloids of Mg-calcite that are cemented together by Mg-calcite micrite or stubby spar (dark) on the left, and cemented by a mesh of aragonite needles or blocky aragonite (light) on the right; note that pores with Mg-calcite cement are smaller than those with the aragonite cement; scale bar 100 μm. (C) Site 7, depth 40 m: a photomicrograph in plane-polarized light of friable *Halimeda* grainstone to packstone that has a 'house of-cards' structure with the voids partially to completely filled with layers of fine-grained skeletal sand. The basal part of each filling is cemented by aragonite [A] and the upper part is cemented by very thin fringes of Mg-calcite micrite [M]; scale bar 0·2 cm. (D) Site 1, depth 110 m: a photomicrograph in plane-polarized light of well cemented *Halimeda* grainstone in which the shelter voids are filled with skeletal sand at the base and silt at the top; the basal portions are cemented by aragonite [A, light grey] and the tops are cemented by Mg-calcite micrite [M, dark grey]; scale bar 0·5 cm. (E) Site 1, depth 110 m: a photomicrograph in plane-polarized light of *Halimeda* grainstone illustrating the wide variation in cementation from no cement at all [N] to areas cemented by a mesh of aragonite needles [A] to areas cemented by fringes of bladed Mg-calcite spar [M]; scale bar 0·5 cm. (F) Site 1, depth 110 m: a photomicrograph in plane-polarized light of medium to coarse-grained grainstone *without* abundant *Halimeda* plates. Most grains are barely cemented by a thin fringe of aragonite needles; scale bar 0·5 cm.

Grainstones, composed mostly of *Halimeda* and other skeletal grains, but with little or no silt (and homogenous in composition), contain very little aragonite cement and grains are commonly cemented by rinds of bladed and fibrous Mg-calcite. What little aragonite mesh cement there is in these sediments occurs as patches surrounded by areas cemented by Mg-calcite, especially when the sediments have obviously been bioturbated. The intimate association of aragonite and Mg-calcite cement indicates that precipitation of these two minerals is coeval and their distribution suggests that one of the major factors controlling mineralogy is the texture and pore size of the sediment. Whatever the other controls, they are subtle.

The sequences of cements described above apply to the first generations of sedimentation or sediment infill between *Halimeda* plates. Subsequent generations of internal sediments, however, filter into the various cavities remaining after the first stage of internal sedimentation and cementation. These cavities include interparticle pores both empty or lined with rinds of cement, and the open tops of larger cavities. These holes are filled largely by silt, or occasionally sand if the cavities are large enough. Like the earlier generation of silt these later sediments are cemented by Mg-calcite micrite or stubby bladed spar. Sand sized internal sediments are almost never cemented with aragonite but by isopachous fringes of bladed and especially fibrous Mg-calcite spar.

Many rocks illustrate several generations of internal sedimentation and cementation in growth cavities. In most instances the deposition of internal sediments is geopetal, so that most cavities are not quite filled to the top with sediment. The multiple generations of cementation within the internal sediments are often recorded on the top of the cavity as rinds of cement several generations thick (Fig. 6-21A).

The packing of silt-sized peloids near the top of many cavity fillings, both in first and second generation fillings, is often so loose that either they must have been pushed apart by the growth of cement or the grains fell into the cavity while cement was being precipitated.

In summary, the mineralogy and morphology of the cements appears to be related to the texture of the internal sediment; because the texture of the internal sediments is highly variable, as a result of small-scale variations in sedimentation and bioturbation, the resulting rocks are compound and complex. Most specimens show an intimate, often quite irregular pattern of the principal types of cement.

Mudstones

Mud not only fills the upper parts of small shelter cavities between *Halimeda* plates but also occurs, more distinctively, as mottled and laminated fine-grained internal sediment between corals in large growth cavities (Figs 6-14, 6-15). These cavities range in size from less than a centimetre wide to over 20 cm across, measured, and in some cases only fragments of material from obviously much larger cavities were recovered.

The sediment is about two-thirds silt composed of Mg-calcite micrite peloids, irregular chips of aragonite produced by boring sponges, and a minor amount of varied skeletal debris. Fine to very fine-grained sand, a minor component of these sediments, is composed of whole to fragmented skeletons.

The laminated mudstones grade in structure from flat-lying geopetal fillings to mammilary or domal stromatolites. The texture ranges from grain-supported to

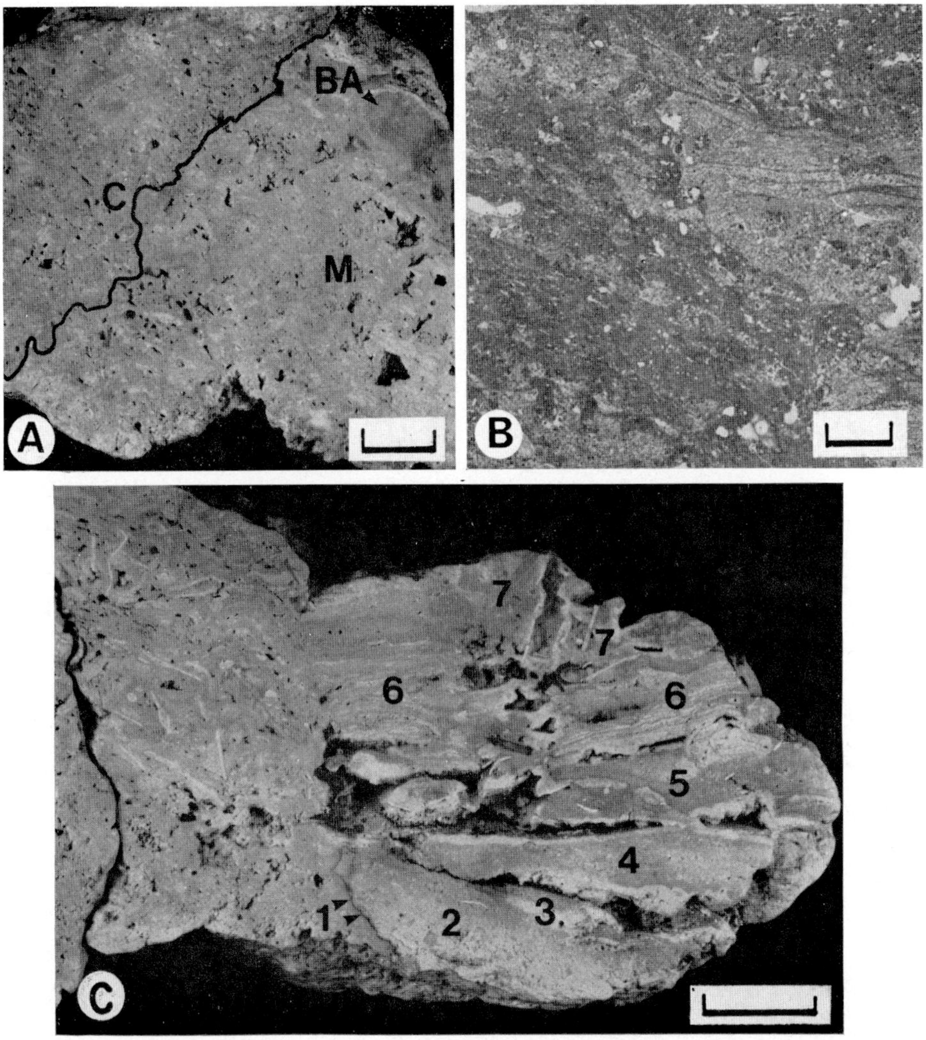

Fig. 6-14. (A) Site 1, depth 110 m: a polished slab of a coral [C] now almost completely altered to mudstone and a cavity filled with mottled mudstone [M] and a growth of botryoidal aragonite [BA]. The complex mottled sediment also contains irregular areas of laminated mudstone; scale bar 1·0 cm. (B) A photomicrograph of the mottled sediment in the cavity illustrated in (A); depicting the mottled mudstone (left) and laminated mudstone (right); the light areas are cemented by aragonite, the dark areas are cemented by Mg-calcite; scale bar 0·2 cm. (C) Site 1, depth 110 m: a polished slab of a compound limestone composed of *Halimeda* mudstone (left) and a cavity filling (right) with numerous stages of sedimentation. A sequence of seven stages can be recognized: (1) an iron oxide coating on the cavity wall, (2) a graded layer of packstone with the top inclined, (3) a thin layer of mudstone, (4) a wedge-shaped layer with a mudstone cap, (5) a thick sediment layer with many laminations, (6) successive laminations cemented by aragonite (light) and Mg-calcite (dark) with a convex aspect, (7) a capping of mud with a few *Halimeda* plates; scale bar 2·0 cm.

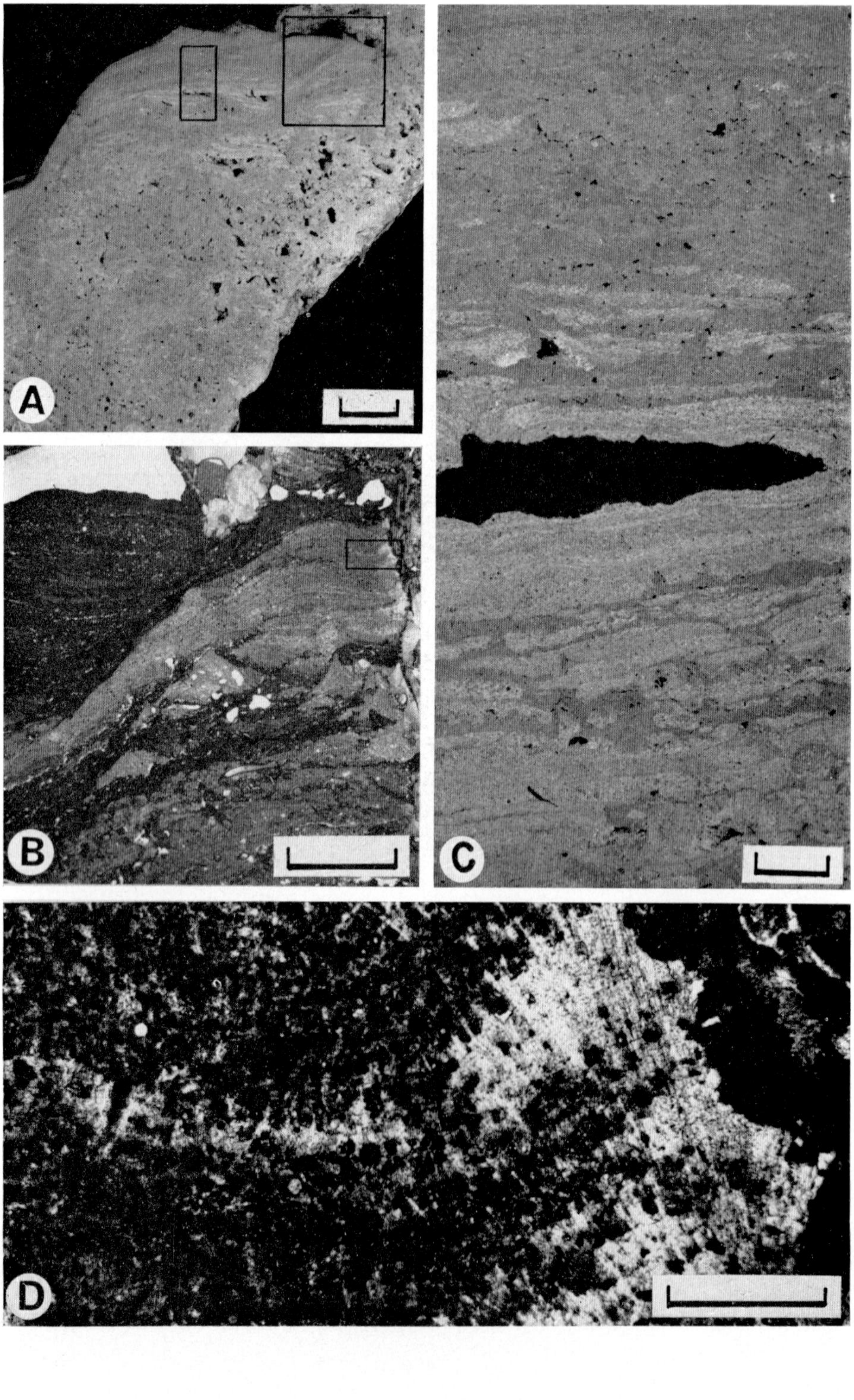
A
B
C
D

cement-supported silts, the cement-supported types are commonly associated with botryoidal aragonite.

Laminated mudstone (Figs 6-14, 6-15)

Most large cavities are characterized by well laminated mudstone, with laminae easily visible both in slabs and in thin sections. The laminations are the result of small differences in grain size, grain packing and type of cement; many of them are graded; their attitude is rarely horizontal and more often slightly irregular. The light and dark coloured laminae seen in hand specimen correspond to aragonite cemented and Mg-calcite cemented layers respectively (Fig. 6-15).

Those laminae cemented by Mg-calcite range from 0·25 to 2·0 mm thick (Fig. 6-16). Each layer has a basal sediment of very fine sand or coarse silt-sized grains, which grade upwards into fine silt and finally to closely packed fine silt and is capped by micrite. The basal part of each lamination is poorly cemented with fringes of 'stubby' blades of Mg-calcite. Cementation is more complete upwards and the top of each lamination is well cemented. The presence of large pores between partially cemented grains, irregularly distributed along the base of each layer, gives a characteristic 'fenestrate' fabric (Fig. 6-16A). The micrite that caps each lamina may be a cement or sediment.

Laminations cemented by aragonite range in thickness from 0·25 to 2·0 mm but thicker layers are more abundant. The grading in aragonite cemented laminae is reversed (Fig. 6–16) to that of those cemented by Mg-calcite. In some laminations there is a basal zone of fine sand to coarse silt-sized grains that are often poorly cemented as Mg-calcite cemented laminae. This basal zone is thin and grades rapidly upward into silt well cemented by aragonite 'mesh' cement; this zone is also thin and it, in turn, grades up into the thickest part of the lamination that is composed of a series of closely spaced, predominantly vertically oriented fans comprising a band of fibrous to bladed aragonite. Grains of sediment in this uppermost zone may be abundant or scarce, but are always 'floating' in the cement and not supporting one another (Fig. 6-16B). The very top of the lamination is a series of needles projecting upwards with little or no sediment.

The basal zone of densely packed and cemented sediment is commonly missing, especially in laminated fillings composed of many thin laminations. As a result most laminae are composed only of a basal zone of densely packed and cemented grains that grades up into spherulitic cement with floating grains. Aragonite laminations are commonly separated at intervals by thin laminations of fine silt cemented by Mg-calcite micrite (Fig. 6-16B). Groups of aragonite-cemented laminae often extend laterally into rounded, upward-projecting pinched terminations (Fig. 6-15C) and

Fig. 6-15. Site 1, depth 110 m: a polished slab of a cavity filling composed of mottled mudstone capped by well laminated sediment; the thickening of the sediments to the right indicates that accretion from the left was a persistent feature; the boxes outlined in ink are the areas of (B) and (C) scale bar 1·0 cm: (B) a photomicrograph in plane-polarized light illustrating large areas cemented by aragonite (light grey) and Mg-calcite (dark grey); the box outlined in ink is the area pictured in (D); scale bar 0·5 cm. (C) a photomicrograph in cross-polarized light illustrating silt cemented by aragonite (light grey areas) and Mg-calcite (dark grey areas). The light areas have little sediment and much cement, the dark areas have much sediment and less cement; scale bar 0·2 cm. (D) a photomicrograph in cross-polarized light illustrating the spherulitic growth of aragonite next to the cavity wall grading outward into the cavity where silt is cemented by aragonite; scale bar 500 μm.

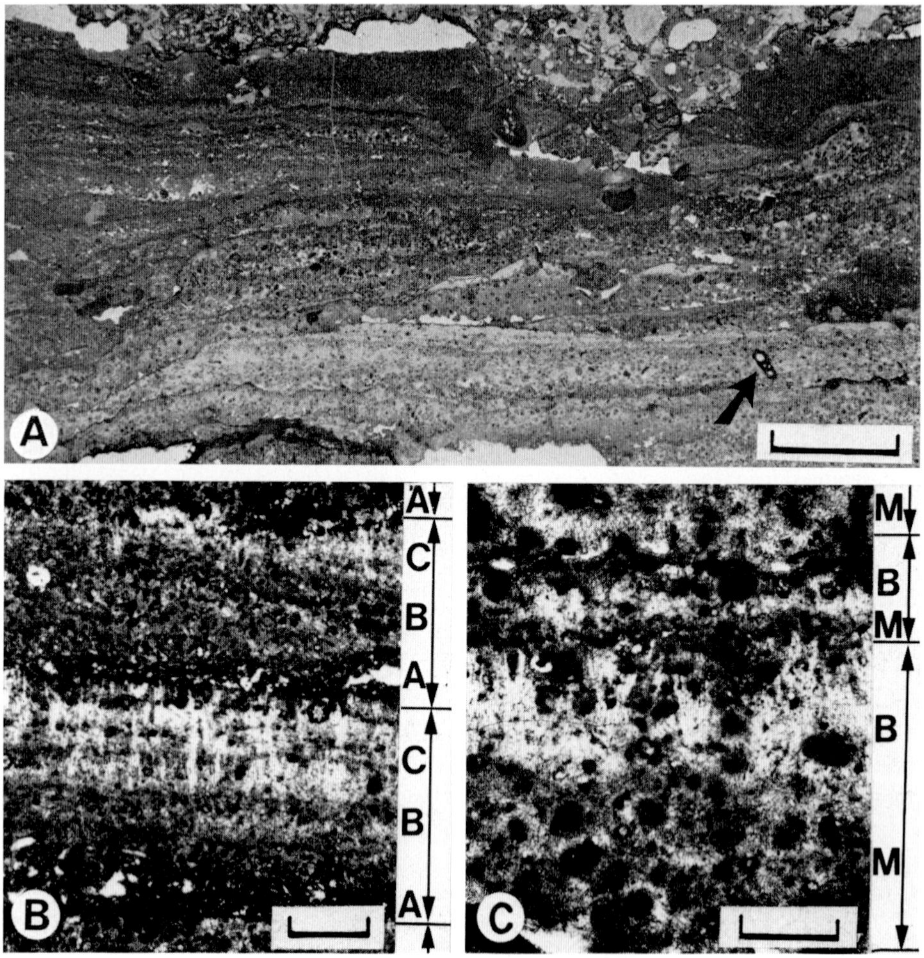

Fig. 6-16. Site 1, depth 67 m: (A) a photomicrograph in plane-polarized light of a cavity filled with silt cemented by aragonite (light areas at base) and Mg-calcite (dark areas at middle and upper left). The Mg-calcite cemented layers have a lower zone cemented by stubby and bladed spar and an upper zone cemented by micrite. The aragonite-cemented layers are cement-supported, suggesting that the cement was precipitating in the cavity at the same time that sediments were filtering in. The presence of whole skeletons such as foraminifers (arrow) indicate that the aragonite is not replacing particles; scale bar 0·5 cm. (B) Site 1, depth 110 m: a photomicrograph in plane-polarized light of layers of silt cemented by aragonite and Mg-calcite. Each layer is composed of a basal zone of silt cemented by Mg-calcite [A] a central zone cemented by a mesh of aragonite needles [B] and an upper zone of spherulitic aragonite cement with floating silt grains [C]; scale bar 500 μm. (C) Site 1, depth 110 m: a photomicrograph in plane-polarized light of laminated silt in which the basal part of each layer is poorly cemented by a mesh of aragonite needles [M] while the upper part is botryoidal aragonite [B] with floating silt grains; scale bar 200 μm.

laminations cemented by Mg-calcite appear to overlie and fill in the resultant cavity.

The laminations cemented by different minerals occur either in irregular alternations (Fig. 6-16A) or as thick sequences of one or the other composition (Figs 6-15C, 6-16A).

Stromatolites

These structures are formed of laminated mudstone, in the form of irregular to superimposed domes, and occur as cavity fillings (Fig. 6-17). Although rare, they are found either in the mottled sediment or at the gradational contact between mottled and laminated sediment. In slabbed section, domal structures range in size from a few millimetres to a centimetre across and resemble small domal algal stromatolites designated LLH (Logan, Rezak & Ginsburg, 1964) (Fig. 6-17). Domes may be truly convex to bell-shaped and may be singular to repetitive, forming crude columns (Fig. 6-17A). In several instances the domes form a series of parallel columns.

In thin section the structures are composed principally of laminated silt, are *always* cemented by aragonite, and the sediment between domes is cemented by Mg-calcite, commonly micrite (Fig. 6-17B). The dark irregular zones seen in slabs are silt cemented by Mg-calcite that extends from the surrounding sediment and drapes over the underlying or arched laminations. The attitude of the aragonite-cemented laminations varies from a gently convex form, in which aragonite mesh cement prevails, to pronounced domal structures cemented by layers of mesh cement that grades into spherulitic aragonite with floating grains. The grains in the surrounding Mg-calcite cemented sediments are more closely packed than those in the aragonite cemented stromatolite. The margins of these stromatolites may be either sharp or gradational, often sharp on one side and gradational on the other; the periphery is often poorly cemented.

Mudstone and botryoidal aragonite

The botryoidal aragonite described above (page 117) often occurs as part of the cavity filling, commonly intimately associated with the fine grained sediment. In some examples aragonite arrays grade laterally into aragonite cemented laminae (Fig. 6-15D). When cavities contain only aragonite mammilons, those that grow upward from the cavity floor contain abundant sediment embedded in the cement (Fig. 6-17C, D). Such basal growths start out from a mesh cement but quickly develop a spherulitic texture, and scattered through the spherulites are sediment grains, randomly distributed or in crude laminations. In some spherulites the fans are not large and contain many grains (Fig. 6-17D) while in others there are few grains and the spherulites are quite large (Fig. 6-5B). Coalescing spherulites are often separated by a thin rind of Mg-calcite cement or laminae of silt cemented by Mg-calcite micrite.

Mottled mudstone

Mottled sediments occur in the lower parts of many cavity fillings (Fig. 6-15A). The mottling is caused by areas of well packed silt, which is well cemented by Mg-calcite micrite or stubby spar, alternating with areas of loosely packed silt generally poorly cemented by aragonite. Aragonite cemented areas give the impression of being

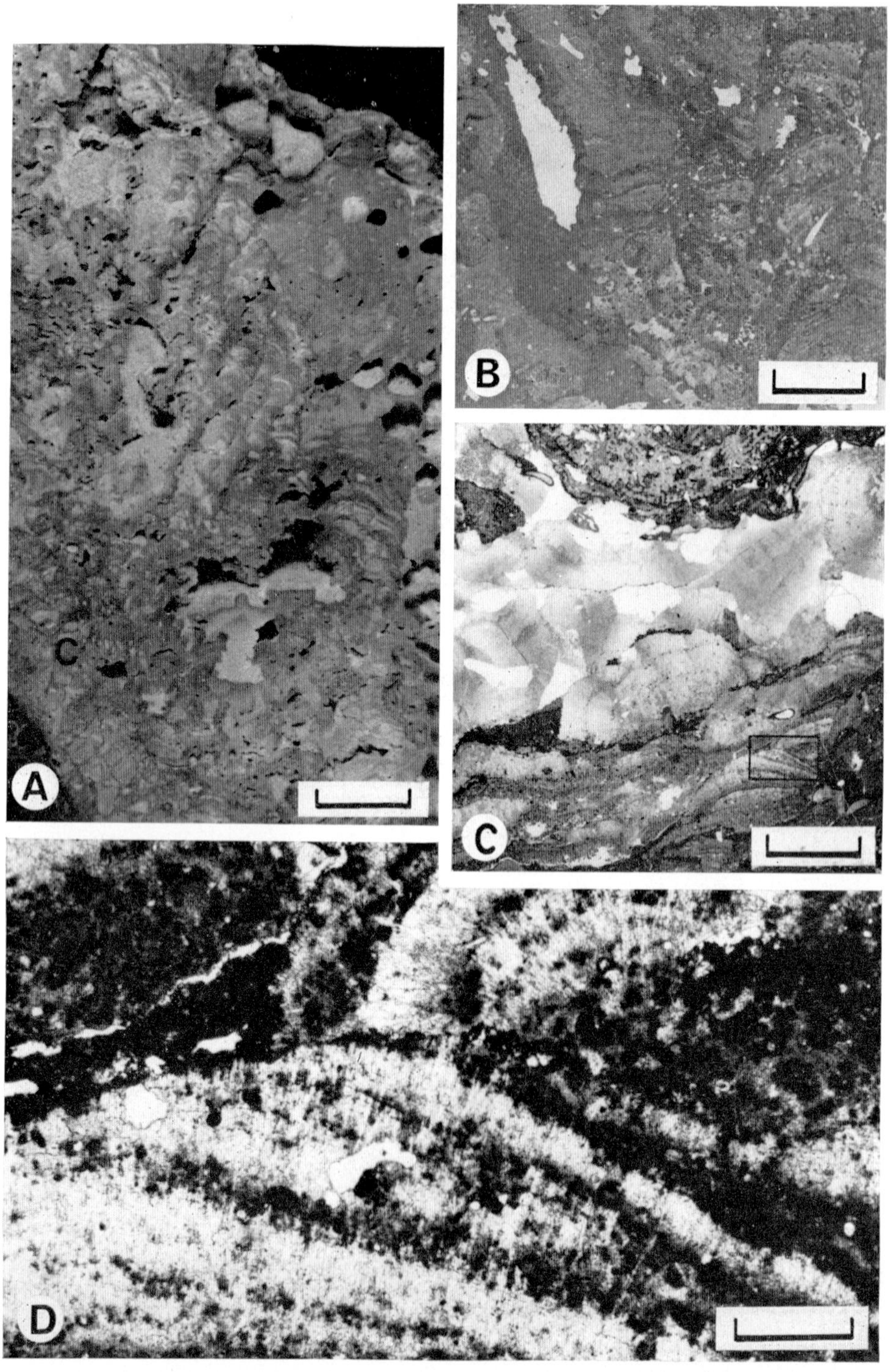
A
C
B
C
D

slightly coarser grained with fewer very fine particles. The contact between mottles varies from gradational to sharp with well cemented sediments on one side and poorly cemented sediments on the other side. We interpret these mottles to be infaunal burrows, with the loosely packed, aragonite cemented sediment the trace or infill of the path of the oranism.

The interpretation of the mottles as the result of bioturbation is supported by the numerous, small, circular to ovoid tubules. These tubules, particularly abundant in the fine sediments, have an interior diameter of 0·4 mm and an exterior diameter of 0·6 mm. The tubes, 0·1 mm thick, are composed of micrite studded with silt grains. These tubes are commonly seen in cross-section in thin section, but occasionally they are tangential and can be seen as sections of linear tubes. The interiors of these tubes may be empty or filled with either aragonite or Mg-calcite cement or geopetal cemented sediment; the periphery is commonly continuous with the Mg-calcite part of the sediment, suggesting that it too may be related to burrowing.

ALTERATION OF CORAL AND LITHIFIED SEDIMENT

The original skeletal texture of corals and depositional texture of lithified sediments is commonly altered by a combination of biological, depositional and physicochemical processes. The hard skeletons and rocks serve as substrates for endolithic organisms, especially sponges, bivalves and worms, which, during their life cycle excavate a variety of holes of varying sizes and shapes into the substrate. Upon death of the organism, sediments filter into the empty holes, partly or completely filling them. These sediments are rapidly lithified and so form part of a new substrate for another generation of boring organisms. This cycle of cavity formation by boring, sediment infill and rapid cementation, repeated several times, transforms the periphery of many coral skeletons or limestones into a compound mudstone to packstone with some samples composed of as many as seven separate stages of alteration.

Formation of cavities

Almost all cavities are made by boring animals (endoliths). The most abundant endoliths are sponges; worms and bivalves are secondary. Endolithic sponges produce

Fig. 6-17. (A) Site 4, depth 122 m: a polished slab of limestone illustrating microcolumnar stromatolites in a cavity. The bottom of the cavity is made by a platey coral almost completely altered by multiple generations of sponge borings filled with internal sediment and cemented [C]. A group of tiny columnar growths begins near the centre and extends diagonally upwards; isolated columns occur below, right and above left of the parallel microcolumns. All of these columnar growths are buried by fine-grained sediment. A younger generation of partly filled sponge borings forms a band along the right side of the specimen; scale bar 1·0 cm. (B) a photomicrograph in plane polarized light of the microcolumnar stromatolites illustrated in (A). This example illustrates the laminated structure and its variations; the light grey laminated sediment cemented by aragonite, the two types of dark sediment, a clotted infill between and interrupting stromatolites, and homogeneous sediment. The stromatolite at left shows bifurcation and interruption of the left branch; scale bar 0·5 cm. (C) site 1, depth 110 m: a photomicrograph in partially polarized light of a cavity filled with growths of botryoidal aragonite. The growths from the ceiling are clear while those at the base are filled with grains of peloid-rich silt. The basal parts of many mammilons at the bottom of the cavity are composed of laminated silt; the box outlined in ink is the location of (D); scale bar 0·5 cm. (D) photomicrograph in partially polarized light illustrating in a close view the interfering growths of botryoidal aragonite that contain layers rich in peloidal silt; scale bar 500 μm.

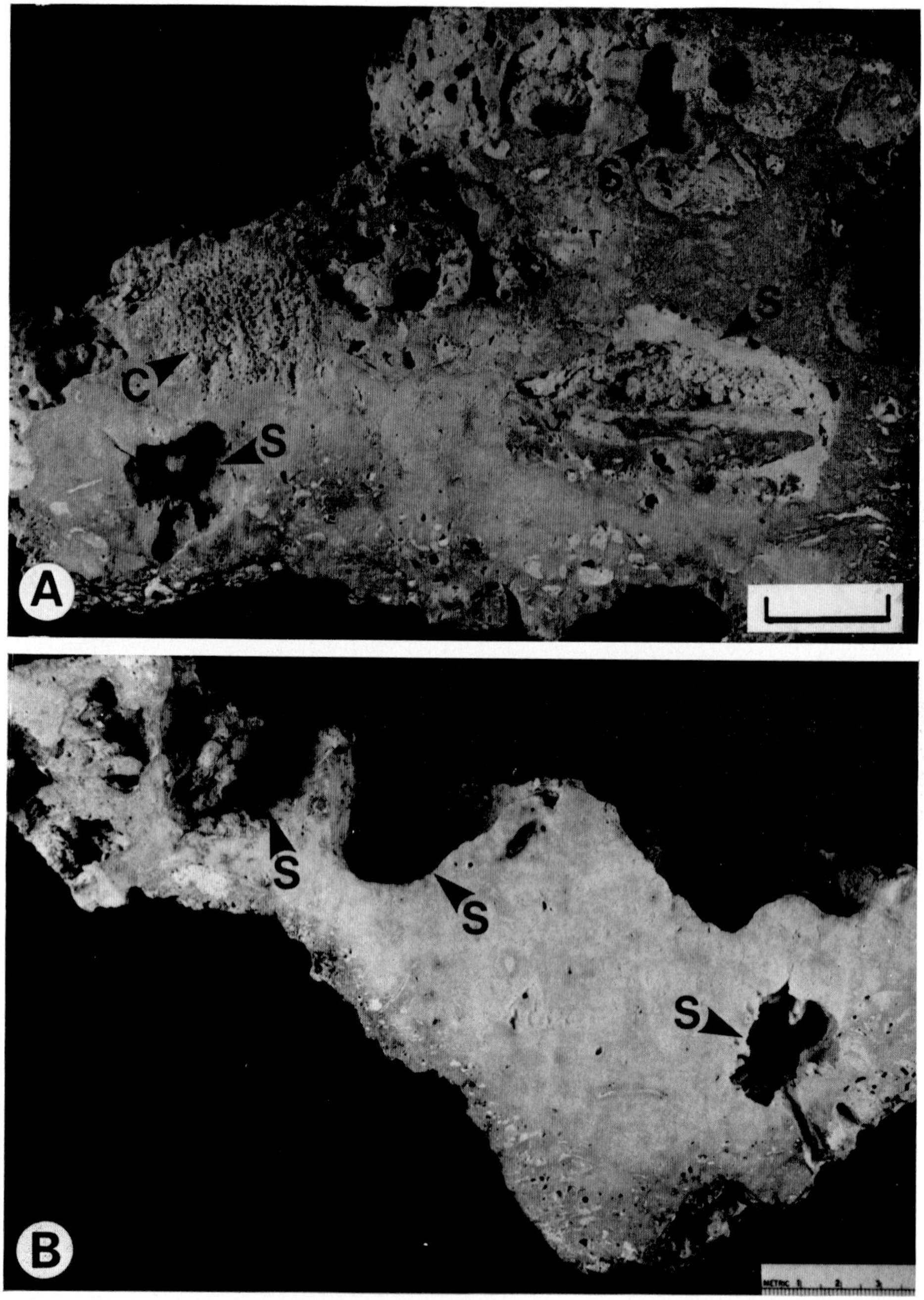

Fig. 6-18. Site 6, depth 105 m: (A) a polished slab of a coral (*Montastraea annularis*) that has been bored by *Siphonodictyon* sponges [S], and *Cliona* sponges [C]; an earlier *Siphonodictyon* boring at right is filled with several generations of internal sediment; scale bar 2·0 cm. (B) a polished slab of a coral (*Montastraea annularis*) which is bored by *Siphonodictyon* sponges, creating irregular cavities [S]; note the excurrent canal leading out from the excavation at lower right; scale bar in cm.

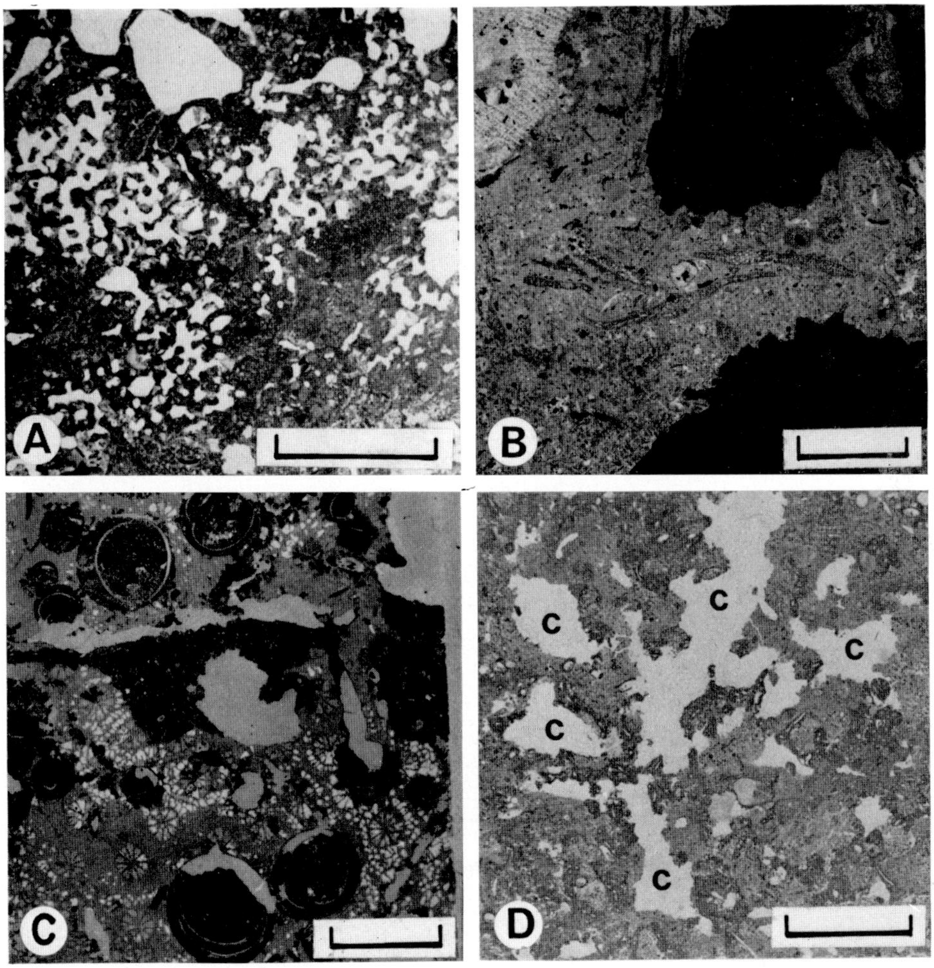

Fig. 6-19. (A) Site 1, depth 143 m: a photomicrograph in plane-polarized light of lithified silt penetrated by the galleries of the boring sponge *Cliona*; scale bar 0·5 cm. (B) Site 1, depth 110 m: a photomicrograph in plane-polarized light of a *Halimeda* packstone with large irregular cavities which we ascribe to the boring by *Siphonodictyon* sponges; scale bar 0·5 cm. (C) Site 6, depth 105 m: a photomicrograph in plane-polarized light of a coral (*Montastraea annularis*) bored by numerous bivalves (*Lithophaga* sp) resulting in many cylindrical holes with the original valves still in place and partially to completely filled with silt; the coral is also bored by *Siphonodictyon* sponges at upper right; scale bar 0·5 cm. (D) Site 1, depth 110 m: a photomicrograph in plane-polarized light of lithified silt with a very irregular cavity [C] formed by an unknown organism or process; scale bar 0·5 cm.

cavities by mining out small chips of the host substrate, generally 20–60 μm in size (Warburton, 1968; Reutzler & Rieger, 1973). Two different types of endolithic sponges are common on the deep fore-reef and their borings are very distinctive: (1) *Siphonodictyon:* large, irregular to ovoid excavations up to 5 cm in diameter, often with the major excurrent canal visible (Figs 6-18 ,6-19B). (2) *Cliona*: innumerable small, ovoid to irregular cavities that, depending on the species, range from 0·4–0·8 mm to 1·0–5·0 mm in diameter (Figs 6-18A, 6-19A). *Cliona* borings are localized to the outer several centimetres of the rock surface but *Siphonodictyon* excavates its cavity as much as 10 cm into the substrate.

Boring worms, most likely polychaetes, form slender to elongate sinuous holes in the rock but are not significant in terms of overall alteration of the rock. Bivalve borings, although not common, are important because they are large (Fig. 6-19C). The commonest endolithic bivalve is *Lithophaga* sp., which creates cylindrical cavities with rounded ends from 2 to 5 mm in length and 0·5–2·0 mm in diameter.

Although most of the cavities are clearly formed by one of the above organisms a small proportion of the cavities have no consistent shape, quite irregular boundaries, and clearly cut previously cemented limestone (Fig. 6-19D). Although boring sponges, especially *Siphonodictyon*, produce a variety of irregular cavities (Fig. 6-18), these irregular cavities (Fig. 6-19D) cannot, with certainty, be ascribed to any particular animal; their irregular borders resemble those produced by solution of limestone exposed subaerially.

Cavity fillings

Cavities are filled with all conceivable permutations and combinations of sediments and cements. Such a wide range of cavity fillings results in a complex series of textures and structures.

Sediment

Sediments in original and secondary cavities are the same as those around and between corals described above. The grain size and thus the composition of these internal sediments is governed by the size of the cavity itself; fine-grained sediment occurs in all cavities, but coarse-grained sediment is restricted to larger cavities. Silt is the commonest internal cavity-filling sediment. Among the earliest cavities to be filled are the intraskeletal pores of coral skeletons and clionid borings (Fig. 6-20A, B, C). Both are commonly filled with silt or cement and often show geopetal arrangement (Fig. 6-20B). These small pores and cavities are filled with only one or possibly two generations of silt. Larger cavities, like those excavated by *Siphonodictyon*, are filled with multiple generations of silt of varying composition or sand (Fig. 6-20D).

Fig. 6-20. (A) Site 7, depth 174 m: a polished slab of coral and sediment; the original pores of the coral (*Montastraea annularis*) are partially filled with fine-grained silt (arrows); scale bar 2·0 cm. (B) Site 1, depth 143 m: a photomicrograph in plane-polarized light of a portion of the coral colony illustrated in (A), depicting the original coral pores partially to completely filled with silt. Some of the pores are instead lined with Mg-calcite bladed calcite spar cement (dark rinds); note the large cavity excavated by a boring sponge (left centre) and partially filled with silt; scale bar 0·5 cm. (C) Site 7, depth 174 m: a photomicrograph in cross-polarized light of the numerous holes excavated by a *Cliona* sponge into a coral colony (*Montastraea annularis*) and partially filled with silt; scale bar 0·5 cm. (D) Site 4, depth 122 m: a photomicrograph in plane-polarized light of a coral (upper left) that has been bored by a *Siphonodictyon* sponge (lower left and right) and the cavities filled with laminated silt. The lower part of the filling at left is inclined and cemented by aragonite, the upper part of the filling is more horizontal and cemented by Mg-calcite micrite; scale bar 0·5 cm. (E) Site 7, depth 125 m: a photomicrograph in plane-polarized light of lithified silt (dark grey) that has been intensively bored by *Cliona* sponges and the resultant holes are now filled with Mg-calcite spar cement; scale bar 0·2 cm. (F) Site 1, depth 110 m: a photomicrograph in plane-polarized light of lithified sand and silt that has been bored by a *Cliona* sponge and the adjacent cavities are now filled with different cements; botryoidal aragonite [B] or several layers of Mg-calcite bladed spar [M]; scale bar 1·0 mm.

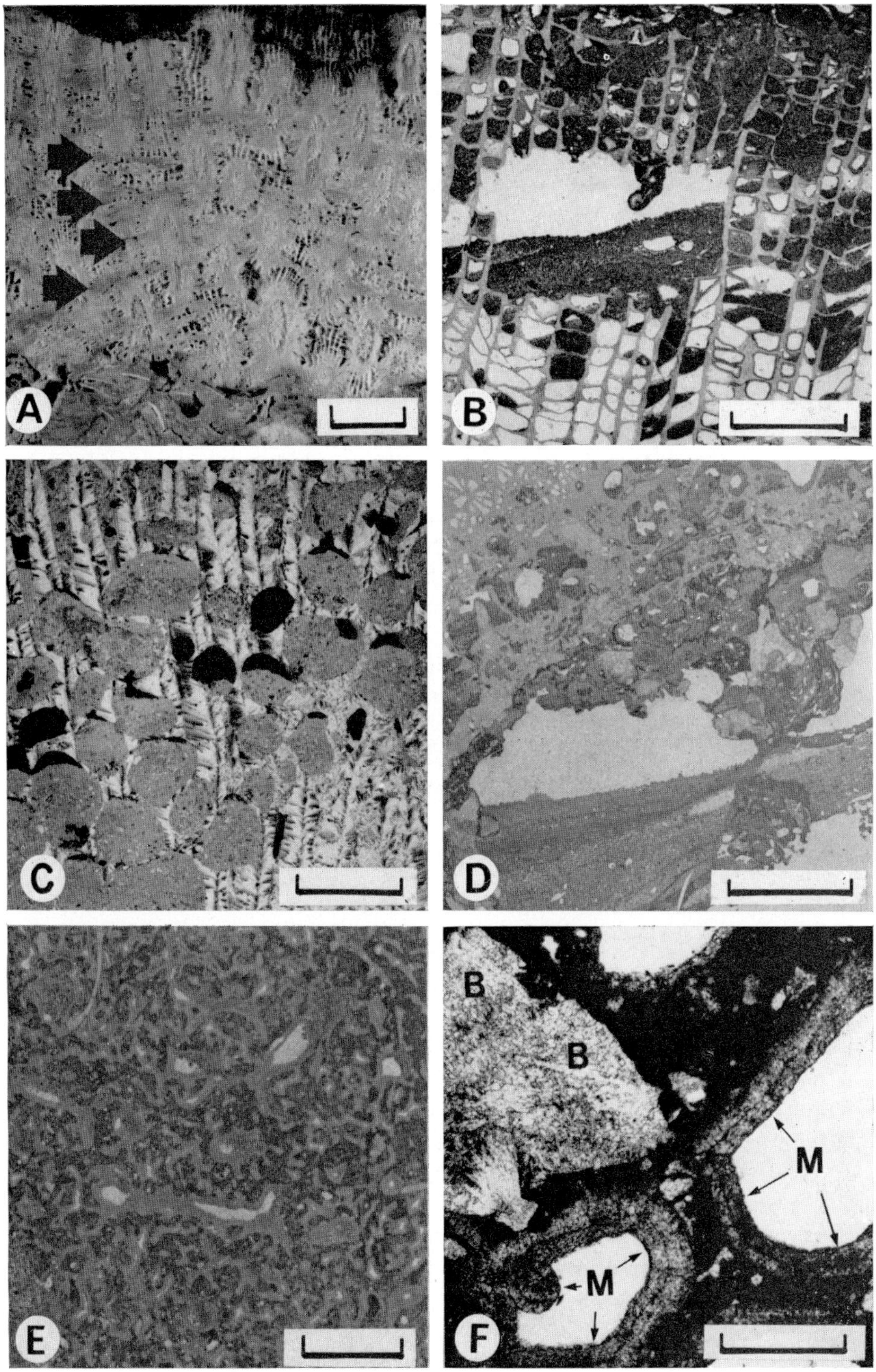
A
B
C
D
E
F
B
B
M
M

Cement

Carbonate cements achieve their most spectacular development in intraskeletal pores of coral skeletons and cavities formed by endoliths. Intergranular cements may range from Mg-calcite micrite to botryoidal aragonite and fibrous Mg-calcite, but cements lining or filling cavities are either thick rinds of Mg-calcite or botryoidal aragonite.

The walls of original coral pores, if devoid of sediment, are rimmed either with rinds of bladed Mg-calcite or asicular aragonite (Fig. 6-20B), often growing in optical continuity with the host skeleton.

Small *Cliona* borings in cemented sediments or in coral rarely contain sediment, but often contain cement. This cement is either rims of bladed Mg-calcite up to 50 μm thick (Fig. 6-20F) or botryoidal aragonite. Clionid borings into lithified sediment that are rimmed with Mg-calcite give a fabric that superficially resembles the leached skeleton of a coral (Fig. 6-20E). Aragonite cement in borings does not grow in optical continuity with the coral substrate as it does in intraskeletal pores.

Larger *Siphonodictyon* cavities are commonly filled with silt and the grains are cemented together by Mg-calcite and/or aragonite (Fig. 6-20D). If both cements are intergranular then aragonite mesh cement is invariably at the base of the cavity filling and extends some distance up into the sediment, sometimes almost to the top. The rest of the intergranular cement to the top of the cavity filling is Mg-calcite, generally as stubby bladed spar. Most commonly the upper few layers in the cavity filling are cemented by well developed Mg-calcite spar and the cement extends out into the open part of the void and around the upper part of the cavity. If sponge borings cut sediment grains of aragonite composition then botryoidal aragonite commonly develops from these aragonite grains.

Most large cavities as well as many *Cliona* borings, because of their size, have multiple generations of cement and sediment. One distinctive style is laminated silt cemented alternately by aragonite and Mg-calcite; other cavity fillings are mottled; and if the cavity filling is incomplete, the space above the geopetal may be filled with botryoidal aragonite (Fig. 6-5B).

A charateristic cavity filling is the partial infill of internal silt with a rim of cement extending around the roof of the cavity. This first generation fill is often followed by a second and a third generation until the base of the cavity is filled by alternating layers of cemented sediment and the upper part of the cavity is rimmed by successive generations of cement (Fig. 6-21A).

Alteration of reef-building corals

Completely unaltered reef-building corals are rare in the limestones. All or part of most corals are altered to some degree to mudstone. The alteration is produced by one or more generations of cavity filling. The first step is infill of skeletal pores and cavities, most commonly along the periphery of the skeleton, where sediment can easily filter in from the surface. When pores are relatively large and interconnected such as *M. cavernosa*, sediment can filter deep into the skeleton. The second step is cavity formation by endoliths, especially sponges. Clionid borings are limited to the peripheries of the coral skeletons, but borings made by *Siphonodictyon* or *Lithophaga* extend deep into the skeleton.

Thin plate-like corals, such as *Agaricia sp.* and some specimens of *M. annularis*,

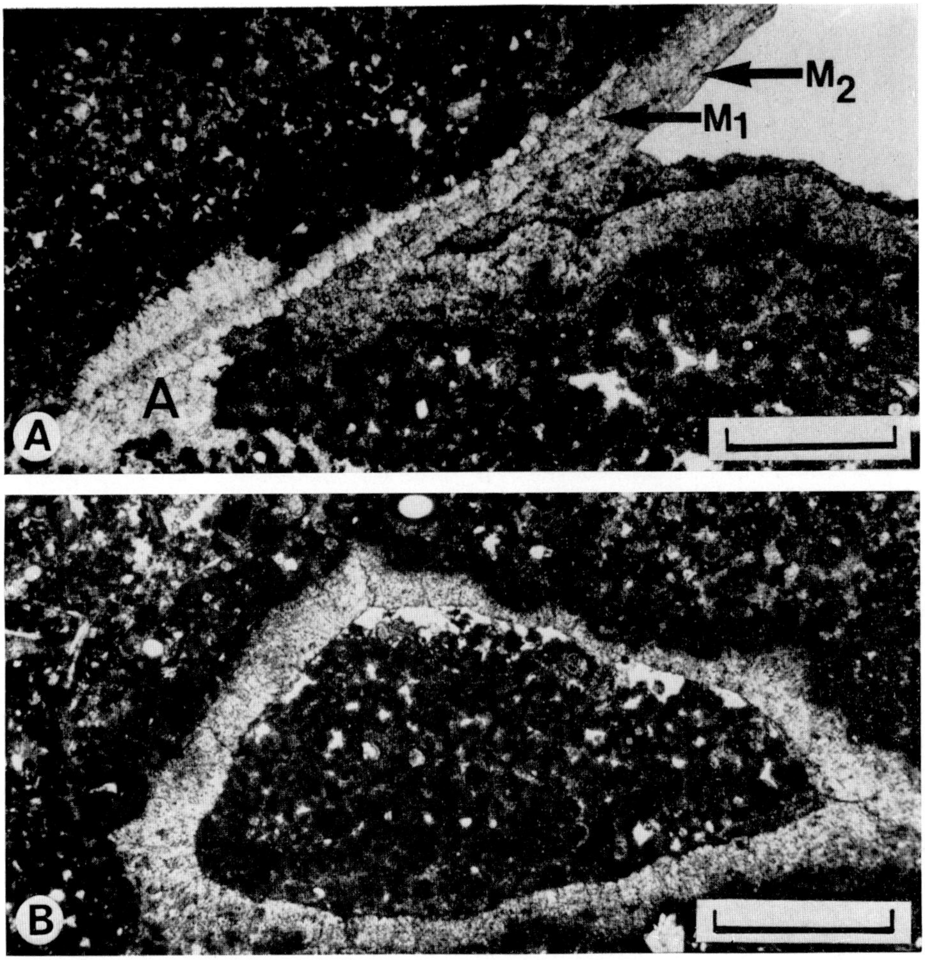

Fig. 6-21. Site 1, (A) depth 110 m: a photomicrograph in plane-polarized light of a peloid-rich mudstone with a cavity (right) that is filled with layers of cement and sediment almost identical to the host rock. The cavity roof runs from lower left to upper right and three separate stages of cement precipitation can be recognized both on the roof and in the sediments: a first stage of aragonite [A], a second stage of Mg-calcite [M1] and a third stage of Mg-calcite [M2]; scale bar 500 μm. (B) Depth 88 m: a photomicrograph in plane-polarized light of a peloid-rich mudstone to wackestone into which a cavity has been excavated and later lined with Mg-calcite spar. Peloid-rich silt was deposited in the cavity after the cement had precipitated; scale bar 500 μm.

are particularly susceptible to alteration because both *Cliona* and *Siphonodictyon* can penetrate the entire thickness of the skeleton (Fig. 6–22A, B). In most instances, however, plates of *M. annularis* are thick enough so that only the periphery is affected.

Peripheral alteration of platey corals occurs through repeated generations of boring *Cliona* sponges, which only reach several centimetres into the coral. In such instances the coral skeleton grades outwards to a mottled mudstone with scattered relicts of coral skeleton (Fig. 6–22). In this transition several generations of infill of different generations of sponge borings can be seen. Peripheral alteration may occur

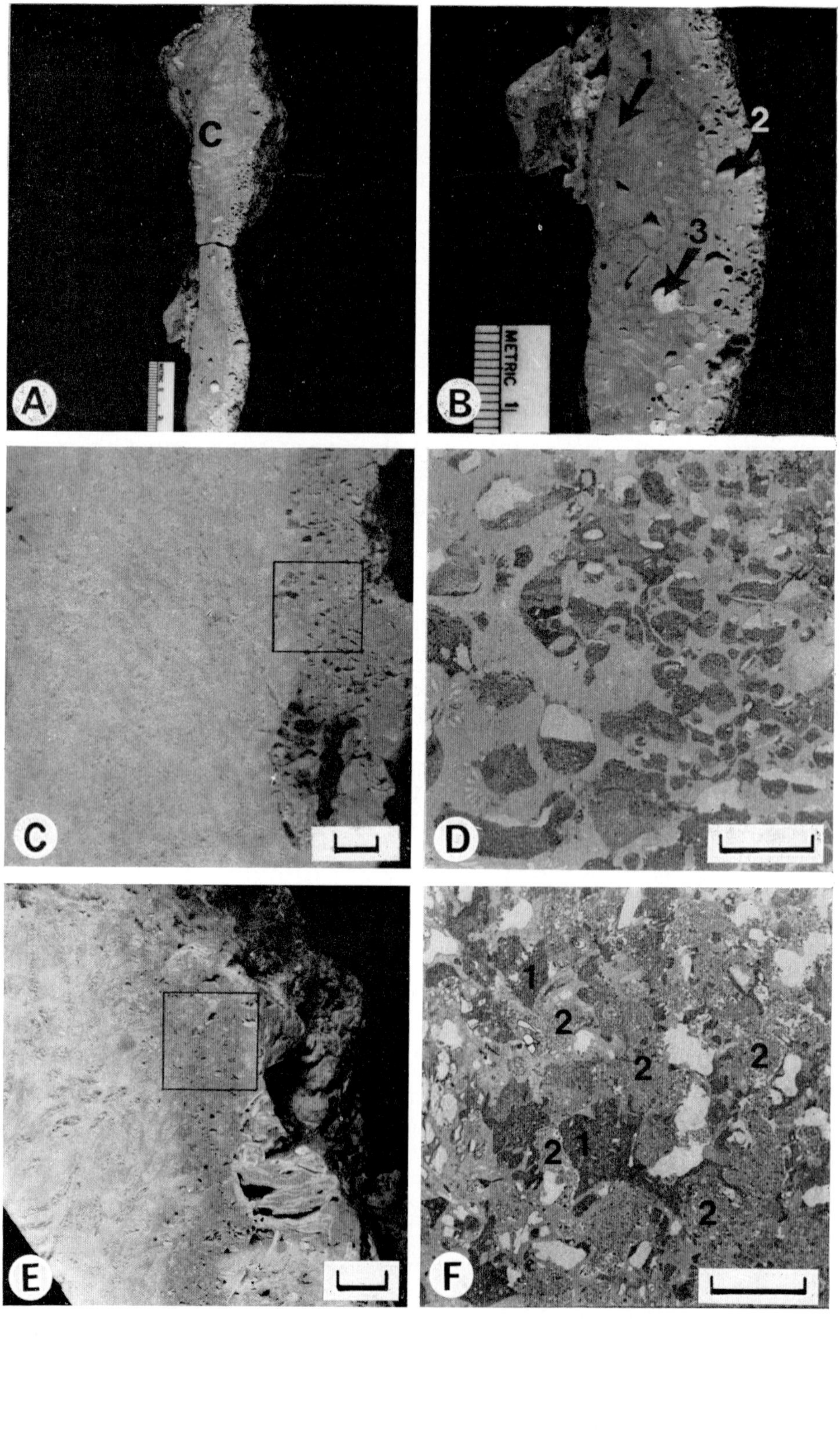
A
B
C
D
E
F
1
2
3
METRIC

only along one side of the coral where it faced an open void, or the entire coral may be altered to mudstone, often to a depth of several centimetres.

Although peripheral alteration, primarily by the combination of clionid sponge borings, internal sedimentation and cementation, may significantly alter coral skeletons, the effects of *Siphonodictyon* borings produce a much more complete modification because they penetrate the entire skeleton. These cavities that reach deeply into the skeleton are often filled with several generations of sediment, thus altering not only

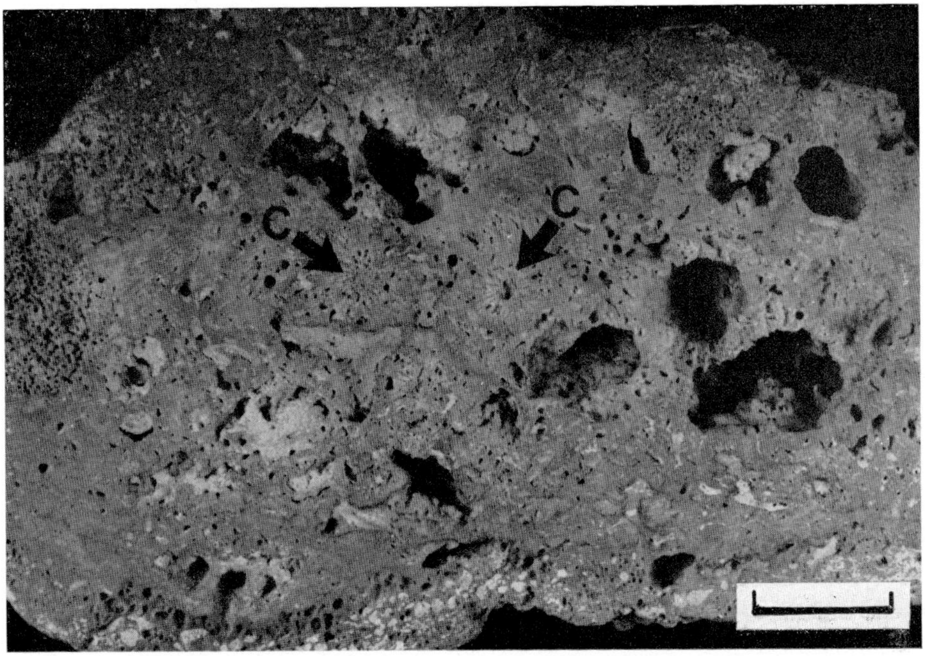

Fig. 6-23. Site 1, depth 110 m: a polished slab of a *Montastraea annularis* colony which has been extensively bored throughout and altered to mudstone by multiple generations of sponge boring, sediment infill and lithification; some of the only remaining evidence of the original colony is in the sediment-filled corallites (C); scale bar 2·0 cm.

Fig. 6-22. (A) Site 1, depth 110 m: a cross-section of a plate-like colony of *Agaricia* polished to illustrate intensive alteration. The colony has been altered by several generations of boring, sediment infill and cementation to form mudstone; the only remaining portion of the coral is around the area C; scale in cm. (B) The lower part of the coral illustrated in (A) in which at least three stages of alteration (numbered in sequence) can be recognized; the remaining coral is at lower left; scale in cm. (C) Site 1, depth 143 m: a polished slab of coral (a columnar colony of *Montastraea annularis*) in which the right margin has been altered to mudstone; box outlined in ink is the area pictured in (D); scale bar 1·0 cm. (D) A photomicrograph in plane-polarized light of the area outlined in (C). The unaltered coral (left) grades into a zone with numerous *Cliona* sponge borings, most of which are partially to completely filled with silt (right); scale bar 0·5 cm. (E) Site 1, depth 110 m: a polished slab of a coral colony, massive *Montastraea annularis*, in which the right hand side (which in turn forms the wall of a cavity filled with laminated sediment) has been altered to mudstone. The box outlined in ink is the area pictured in (F); scale bar 1·0 cm. (F) a photomicrograph in plane-polarized light of the altered margin of the coral pictured in (E). The coral (left) has been altered by one generation of boring and sediment infill (1) only to be bored again and infilled with another generation of silt (2). The final stage of alteration, open cavities, have either yet to be filled or the uncemented sediment was washed out during sampling; scale bar 0·5 cm.

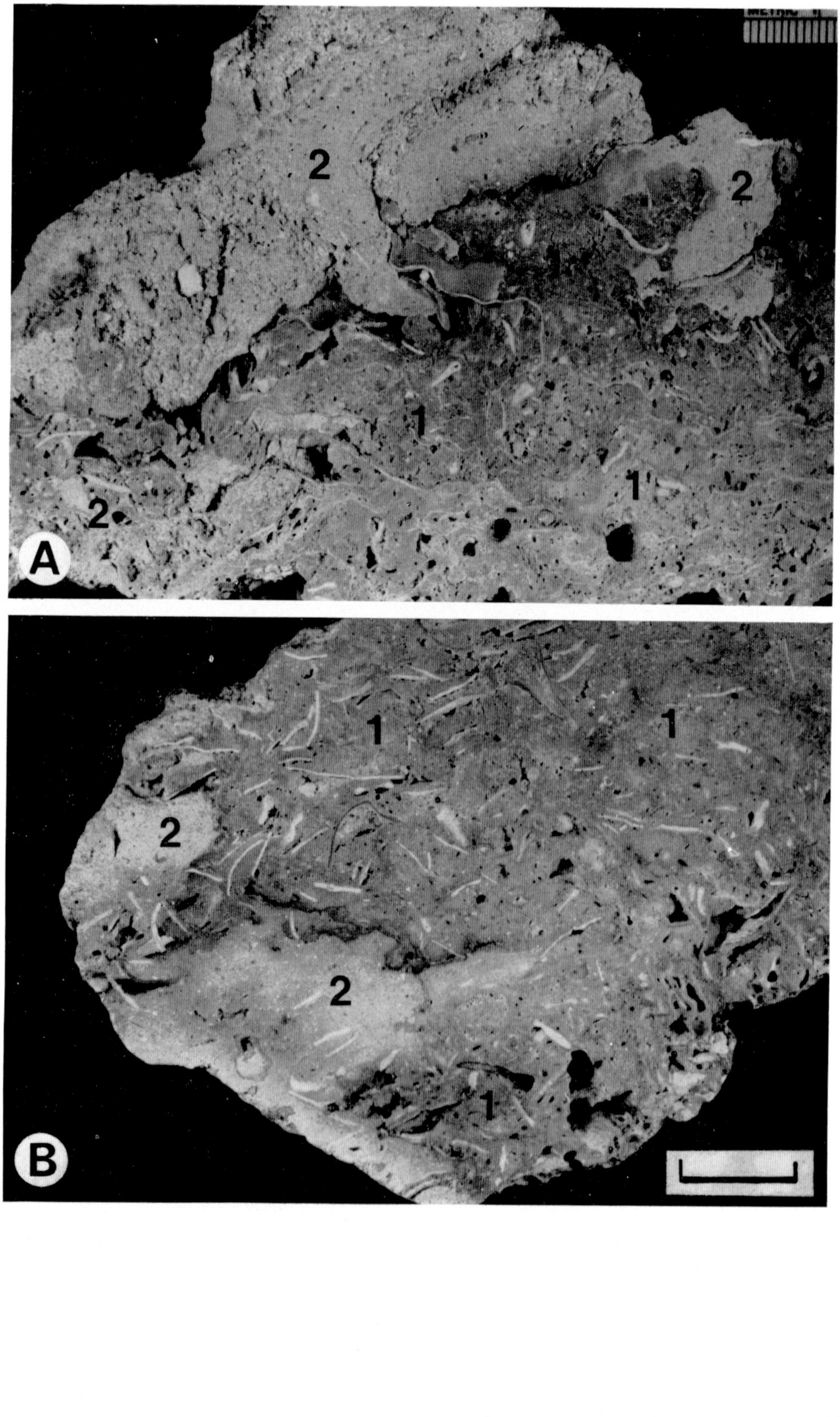
2
2
1
1
2
A
1
1
2
2
1
B

the periphery of the coral but also the centre, and almost completely obliterating the coral structure (Fig. 6-23).

Alteration of lithified sediment

Lithified sediments are altered in the same way as coral skeletons. While the alteration of coral is relatively easy to detect because of obvious differences in structure between skeleton and infilling sediment, original and altered sediment may be similar and thus almost impossible to differentiate (Figs 6-22F, 6-24A, 6-25D). Initially sediments filter into the shelter cavities of partially cemented sediments; subsequent generations of sediment fill new cavities excavated into previously cemented sediment.

Perhaps the commonest style of alteration is peripheral alteration like that in corals described above. Commonly in *Halimeda*-rich sediments there are large irregular cavities formed by species of *Siphonodictyon* sponges filled with lime wackestone to *Halimeda* packstone that is less well lithified and obviously different from the host rock. Although slightly different in colour and degree of lithification the composition of the host sediment and infilling sediment is often quite similar (Fig. 6-24). The combined excavation of cavities by *Siphonodictyon* in the interior of the sediment and the many generations of *Cliona* borings, in addition to irregular cavities formed by unknown agents has clearly transformed many of the rocks into mudstones with an irregular composition and mottled appearance on the scale of centimetres (Figs 6-23, 6-24) and millimetres (Figs 6-22F, 6-25B, C, D). Because many of the limestones examined consist of mixtures of coral and interskeletal sediment, both altered to varying degrees, it is difficult to reconstruct the original proportions of the two components. In some instances the alteration is so pervasive (Fig. 6-26) that the original composition can only be inferred.

Iron–manganese surficial coatings

Limestones from most localities are characterized by coatings of iron–manganese oxides on exposed surfaces and on the surfaces of hidden or buried cavities (Fig. 6-27), giving the surfaces a brick-red or black colour. These coatings were particularly conspicuous in fresh underwater exposures where they stood out against the white limestone and were easily seen from the submersible. The occurrence and abundance of these coatings varies with depth; limestones from less than 40 m deep have no coatings and are buff to white; limestones from the wall and from talus blocks on the sloping fore reef are distinguished by many red-brown iron oxide coatings and occasional black iron–manganese coatings; limestones from the deep ridge and furrow structures off Glovers Reef show a few iron-oxide coatings, but all samples have black iron-manganese coatings.

Analysis of the coatings and small amounts of the underlying limestone could detect no minerals other than the aragonite and Mg-calcite of the host limestone. Analysis for major elements by atomic absorption and specific elemental analysis by

---

Fig. 6-24. (A) Site 1, depth 110 m: a polished slab of *Halimeda* packstone to wackestone (1) that has been bored (upper left) and the cavities filled with sediment of almost identical composition (2); scale in mm. (B) Site 7, depth 125 m: a polished slab of limestone illustrating the alteration of sediment. The original sediment, a *Halimeda* wackestone (1) has been bored (lower left) and the cavity infilled with a later generation of *Halimeda*-rich fine-grained sand (2) of similar composition; scale bar 1·0 cm.

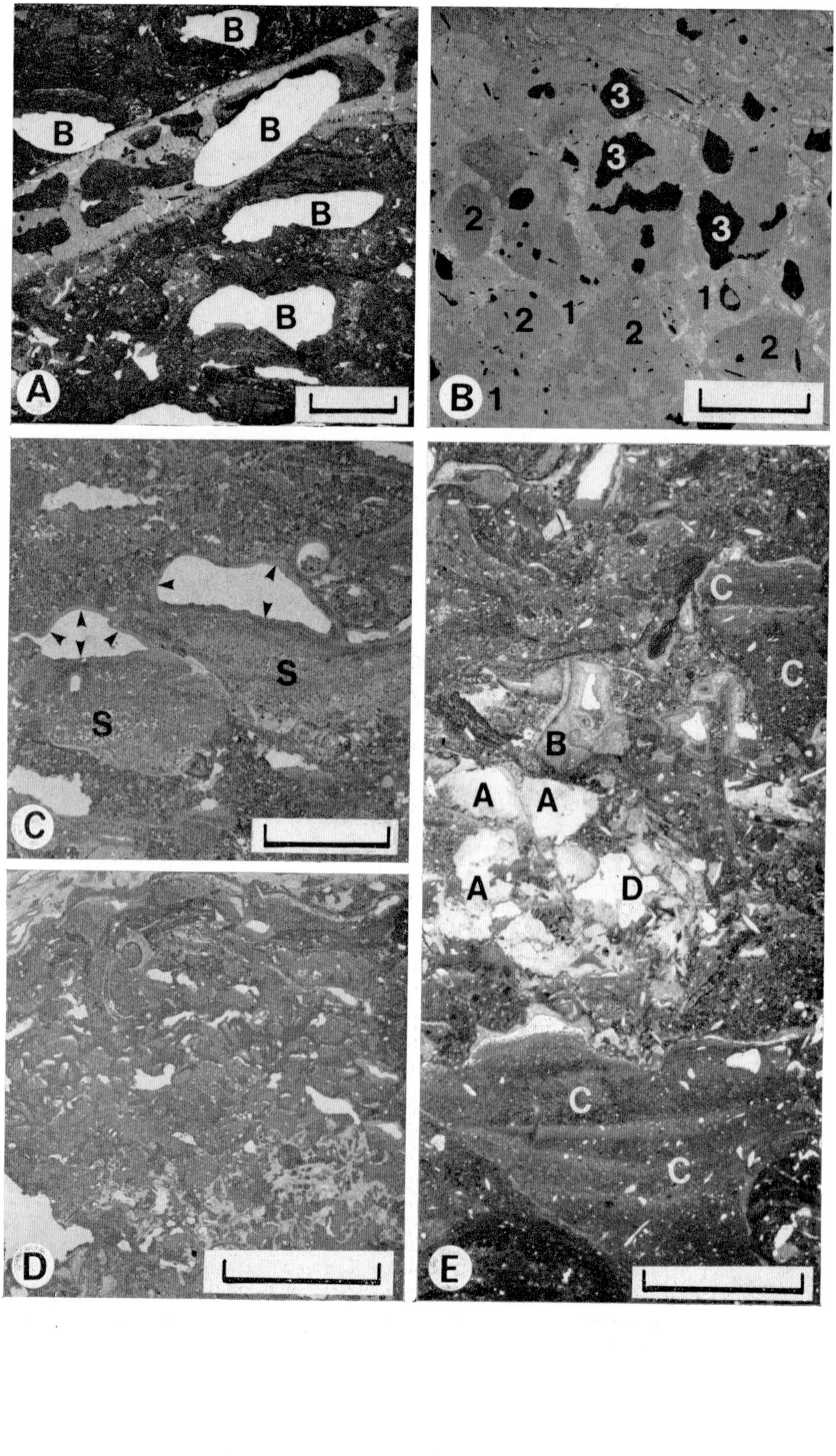
A
B
B
B
B
B
B
3
3
3
2
2
1
1
2
2
1
C
S
S
D
E
C
C
B
A
A
A
D
C
C

microprobe (courtesy L. Land, University of Texas) shows that the brick-red coatings are rich in iron and the black coatings are rich in iron–manganese.

Brick-red iron-oxide rinds

Iron-oxide coatings vary from thin veneers on cavity walls to rinds as much as 0·25 mm thick. The thick rinds are composed of two zones: a dense, opaque outer zone and a diffuse inner zone, which in thin section is transparent and can be seen to impregnate the host limestone. The inner zone is often characterized by a series of small tubules filled with translucent brown material, and extending some 0·25 mm further into the rock. These tubules which range from 20 to 30 μm in diameter and are up to 200 μm in length, are the same size and shape as endolithic algal borings.

Coatings occur on all kinds of surfaces, cavity walls, surfaces of internal sediments and cements. The upper surface of any particular generation of sediment, either *Halimeda* sand or laminated sand and silt deposited as internal sediment, may be veneered with a coating of iron-oxide. Botryoidal aragonite and thick fringes of fibrous Mg-calcite that form around cavity walls are often covered with rinds up to 200 μm thick. Irregular cavities excavated by boring sponges are also often rimmed by brick-red coatings.

The iron-oxide coatings are not restricted to a single generation or event in the history of these rocks, but occur sporadically between generations of cementation and/or sedimentation. A common example of this is a cavity whose walls have been coated with iron-oxides, and then the cavity itself has been filled with sediment, that is in turn cemented (Fig. 6-27B, C). This succession indicates a period of non-deposition after cavity formation. Original sediment forming the host limestone and secondary infilling sediment are often so similar that the coating is the only remaining clue as to the presence of a cavity stage (Fig. 6-27B, C).

Iron-manganese coatings

Black iron-manganese coatings are very similar in structure to the iron-oxide coatings, but thicker. Individual coatings may be as thick as 0·5 mm, with an outer

Fig. 6-25. Site 1, (A) depth 143 m: a photomicrograph in plane-polarized light of a limestone illustrating several stages of boring; first the plate-like coral colony was bored by a sponge after deposition and the sediment filtered into the resultant holes; after the sediment was lithified the rock was bored again by a sponge, creating the irregular, empty cavities; scale bar 0·5 cm. (B) Depth 110 m: a photomicrograph in plane-polarized light of highly altered mudstone. The original mudstone (1) has been bored by sponges and the holes are filled with a mudstone of similar composition (2) and after lithification the rock is bored again (3), but this last generation of holes is not yet filled with sediment; scale bar 0·5 cm. (C) Depth 110 m: a photomicrograph in plane-polarized light of *Halimeda* packstone in which the lithified sediment has been bored by sponges and the resultant cavities are partially infilled with laminated silt [S] and succeeded by a rind of Mg-calcite spar cement (arrows); scale bar 0·5 cm. (D) Depth 143 m: a photomicrograph in plane-polarized light of two corals (top and bottom) with sediment between, that has been altered. The original sediment (mud) is dark grey and the new silt that has filled sponge borings is light grey; scale bar 0·5 cm. (E) Depth 110 m: a photomicrograph in plane-polarized light of a *Halimeda*-rich wackestone that has been extensively bored and the resulting holes filled by a variety of different sediments and cements. Some of the holes are filled by botryoidal aragonite [A], some are filled by layers of Mg-calcite spar [B], some are filled with layers of geopetal silt [C] with the cavity roofs lined with Mg-calcite cement, while some have remained empty [D]. Many cavities show successive generations of sediment infill followed by precipitation of Mg-calcite spar. Because of this extensive alteration the composition of the original sediment is difficult to determine; scale bar 0·5 cm.

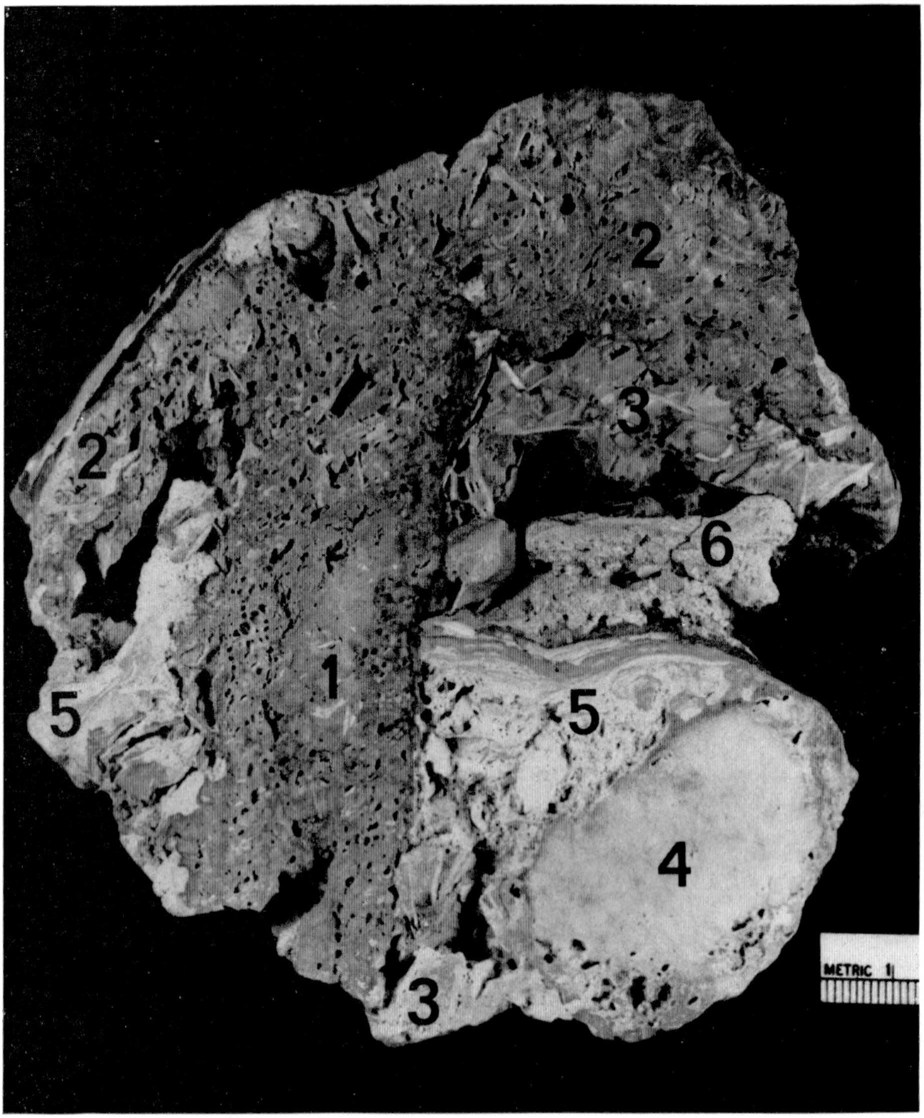

Fig. 6-26. Site 1, depth 110 m: a polished slab of one of the most complex pieces of wall limestone recovered. A total of six separate stages can be recognized as follows: (1) a platy coral extensively altered to mudstone, (2) surrounding *Halimeda* mudstone, (3) boring of the sediment and infill of the resultant cavities with *Halimeda* mudstone, (4) boring and the deposition of a coral fragment in the cavity, (5) partial filling of the cavity with *Halimeda* grainstone grading up into laminated mudstone with an iron-oxide coating at the top indicating a period of non-deposition, (6) a partial filling of grainstone; scale in cm.

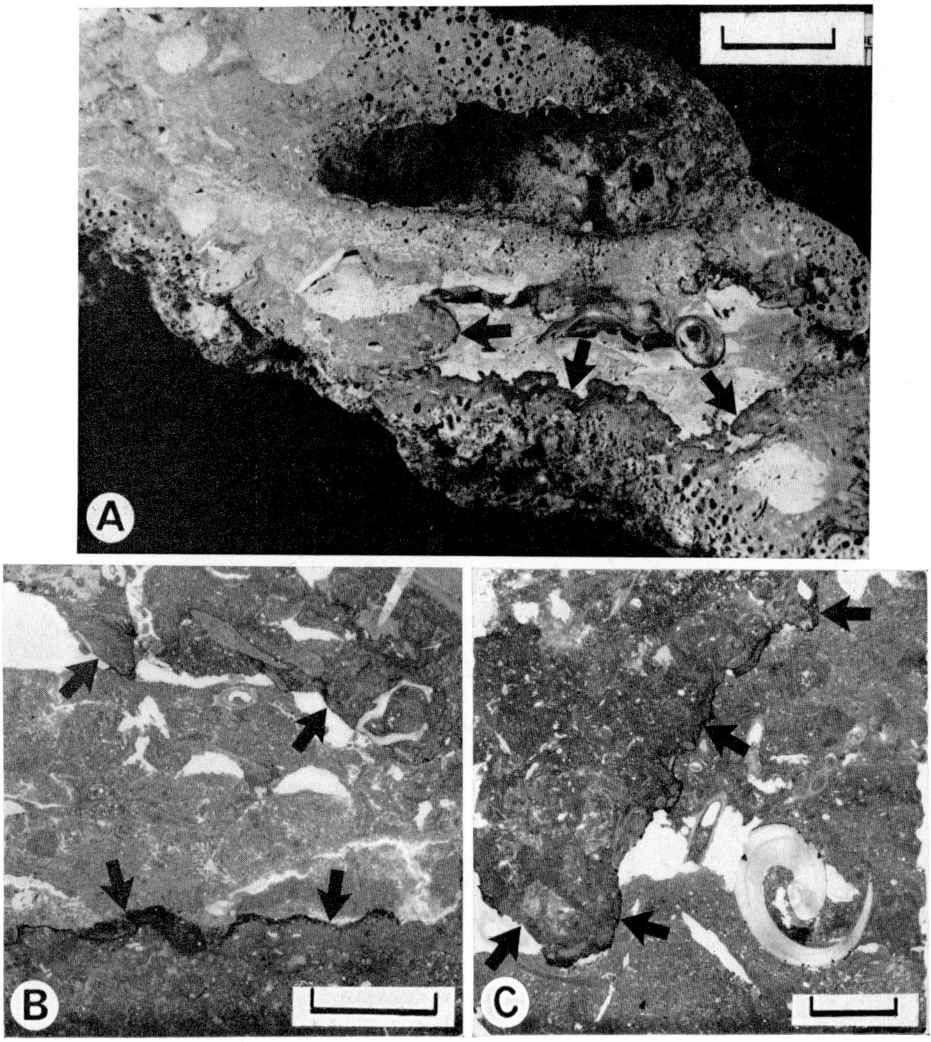

Fig. 6-27. (A) Site 1, depth 88 m: a polished slab of intensely altered mudstone with well developed iron–manganese oxide coatings on cavity walls (arrows); note the empty mollusc boring (top) the margins of which have been bored by a *Cliona* sponge; scale bar 2·0 cm. (B) a photomicrograph in plane-polarized light of a mudstone with a cavity whose walls are coated with iron–manganese oxides (arrows) indicating a time when it was open. The cavity was subsequently filled with sediment; scale bar 500 μm. (C) a photomicrograph in plane-polarized light of that part of the cavity illustrated in (A), depicting the iron–manganese coatings on the cavity walls (arrows). The later fill of sediment in the cavity results in a fabric that resembles stylolites; scale bar 500 μm.

dense zone 0·1 mm thick and an inner transition zone 0·4 mm thick. The zones are, however, irregular, with several smaller, laterally discontinuous zones making up the coating. Tubules penetrating the substrate are also associated with these coatings and are, like those in the iron-oxide coatings, ascribed to the filling of endolithic algal borings.

The black coatings occur in the same situations as the iron-oxide coatings (Fig. 6-27), but they are much more widespread as coatings on surface outcrops in the deeper reaches of the margin, below about 150 m.

SUMMARY

The coral rich limestones along the deep reef margin exhibit all degrees of lithification, but most are well cemented. The limestones are cemented by aragonite and/or Mg-calcite; there is no indication of subaerial exposure, neither loss of metastable carbonate minerals nor the presence of calcite.

Mg-calcite is the commonest cement, occurring primarily in interparticle pore as rinds of micrite or bladed spar. The bladed spar is precipitated as splays. The rarer aragonite develops more varied cement morphologies in the form of a mesh of crystallites between particles, epitaxial overgrowths and botryoidal crystal arrays.

The lithified sediment ranges in composition from *Halimeda* grainstones to finely laminated calcilutites that fill primary and secondary cavities. The composition of these sediments is very similar to unconsolidated reef margin sediments: *Halimeda* plates intermixed with medium to fine-grained sand and silt.

Cementation of *Halimeda* grainstones to wackestones is variable but coarse-grained sediments are generally cemented by aragonite while fine-grained sediments are cemented by Mg-calcite; the distribution of the cements is therefore irregular with definite sequences of cements only present in grainstones. Mudstones, generally deposited as internal sediments, range from well laminated to mottled. Laminated mudstones range from geopetal to stromatolitic. When cemented by aragonite the grains in laminated mudstones are commonly cement-supported.

Both coral and cemented sediments are substrates for endolithic organisms and so have been intensively bored. Because sediment quickly infills cavities that are vacated and is in turn rapidly cemented, new substrates are rapidly created for succeeding generations of endoliths. The process of boring, infill by fine grained sediments, and lithification repeated many times has all but replaced the original depositional fabrics with mottled mudstone.

Chapter 7

Comparative anatomy, organism distribution and Late Quaternary evolution of modern reef margins

INTRODUCTION

The margins of modern reefs and carbonate platforms facing deep water are steep. The precipitous nature of these slopes was recorded from the very earliest studies of reefs summarized by Darwin (1842), who noted the abrupt drop-off just seaward of reefs in many areas. Subsequent and more detailed soundings and profiles from Pacific atolls (David *et al.*, 1904; Emery, Tracey & Ladd, 1954), reefs in the Red Sea (Grabau, 1913), the margin of the Bermuda Pedestal (Stanley & Swift, 1968), the edge of the Bahama Banks (Hess, 1933; Newell & Rigby, 1957; Emiliani, 1965; Zankl & Schroeder, 1972), the north coast of Jamaica (Goreau & Land, 1974), the margin of many Caribbean islands (Macintyre, 1972; D'Anglejan & Mountjoy, 1973) and the edge of the continental shelf off Australia along the Great Barrier Reef (Maxwell, 1968) and off the Sahul Shelf (van Andel & Veevers, 1967) all record a sudden break in slope somewhere between 30 and 100 m, below which there is a submarine cliff.

Before the extent of fluctuations in sea level during the Pleistocene glaciations was known the upper slopes of barrier and atoll reefs were viewed, as was indeed the entire slope, as the trace of shallow-water reef growth during subsidence. The first concrete evidence of accretion to the reef wall at depths below the limit of luxurious coral growth came from the interpretation of the collections off Funafuti Atoll.* There, David *et al.* (1904) recovered numerous flattish specimens of rock from a submarine cliff (reef wall) and inferred that such rocks or corals from shallower depths would become entangled in the profuse growth of octocorals and in turn trap sediment descending from above. These rocks or coral would then '. . . in common with the sediment at their back, become overgrown by deep-sea organisms, particularly by the encrusting variety of *Lithothamnion* until the whole of the fragmental material has been firmly cemented to the face of the submarine cliff' (p.153). David *et al.* (1904, p.154) concluded 'This coral atoll accordingly affords us the remarkable features of a nearly vertical wall, 500–600 feet in depth, largely fragmental (at all events as regards its surface), which is being constantly added to, and which thus enlarges its periphery oceanwards at an extremely slow rate.'

With the subsequent realization of the nature and extent of fluctuations in sea level produced by Pleistocene glaciations, Daly (1910) suggested that the margins of reefs owed their steepness to erosion during low stands of sea level. Although Vaughan

* The collection of these specimens was a remarkable accomplishment. David's assistants and a seaman sampled the reef wall to a depth of 400 m by hand, from a small boat! An eighty-pound chisel was used to break off pieces of the wall that were brought up in a hemp tangle suspended below the chisel (David *et al.*, 1904).

(1919, p. 236) did not agree with Daly entirely, he was inclined to view the reef wall as an erosional feature: 'It seems to me that most of the platforms are of pre-Pleistocene age, and were wave cut and were remodeled around their edges during Pleistocene time; but this is a subject that needs more investigation.'

Our study combined with other recent surveys provides the further investigation that Vaughan requested. Our results confirm and extend the theory of accretion first suggested by David (1904) and from these results we can outline a model for the development of the reef margin. In this chapter we first compare the physiographical elements of the Belize margin with those of other areas; next we review the distribution of reef-building corals and algae in Belize and elsewhere; then we consider the age and environment of limestones from the reef wall and fore-reef; and finally, using all the evidence, we offer an interpretation of the history of development of the margin during a period of rising sea level and by extension a model of accretion during the entire Pleistocene.

PHYSIOGRAPHY

The reef front

Although the slope that leads from the surf zone to the top of the wall, the zone of reef growth, is extremely variable from place to place, it is characterized by two recurring elements: (1) an overall 'stepped' profile consisting of gently dipping slopes or terraces and intervening steep cliffs, and (2) elongate ridges or pinnacles parallel to the trend of the reef that have been called 'submerged' reefs.

Stepped topography

The degree to which the profile is stepped is first a function of the width of the zone; if the zone is wide then this topography is readily discernible, if the profile is narrow then the features are condensed or absent.

The commonest feature, although often not the most obvious, is the step, con-consisting of a steep cliff or riser extending from 20 to 35 or 40 m and a seaward-dipping slope or tread. The cliff usually begins at a depth of 20 m or so, and at its base there is sometimes a small notch (Aldabra, Barnes *et al.*, 1971; Curacao, Focke, 1978). The slope at the base dips seaward from 15 to 60° and isc alled the '35 m terrace' in the Bahamas (Zankl & Schroeder, 1972) and Bermuda (Meischner & Meischner, 1977), the second terrace in Curacao (Focke, 1978), or the 'fore-reef slope' in Jamaica (Goreau & Land, 1974).

Macintyre (1972) has also noted the widespread occurrence of a terrace at 40 m off many Caribbean islands and remarked that it occurs at the same depth as a terrace off Puerto Rico, in the English Channel, and a number of areas in the North Atlantic. Interestingly, Macintyre also notes the common occurrence of another terrace at a depth of *c.* 80 m. This lower level is also the level at which numerous sediments, obviously deposited in very shallow water, have been found (Ginsburg & James, 1974) in various areas of the world.

Above the step there is commonly recognized another relatively gently dipping slope, called the 'Ten Fathom Terrace' (Stoddart, 1969). This feature is recognized around many Pacific atolls (Stoddart, 1969), at 14·5–18 m around Bermuda (Stanley &

Swift, 1968; Meischner & Meischner, 1977), at 18–20 m off the Bahamas (Newell *et al.,* 1951; Zankl & Schroeder, 1972). The terrace is not easily recognized either in Jamaica (Goreau & Land, 1974) or in Belize (this report). Instead the zone is a gently sloping surface to *c.* 15 m (Jamaica) or 20 m (Belize), consisting of sand channels and intervening ridges of coral growth, oriented to the reef trend.

Seaward of the tread of the step, or '35 m terrace', the bottom either becomes progressively steeper with depth until it rolls over to merge with the top of the wall, or there is a distinct break in slope at *c.* 45 m, where the bottom becomes much steeper, just above the top of the wall (the brow).

Submerged reefs

The ridge off South Water Cay (site 2) along the barrier reef that rises from depths of 33 m to within 15 m of the surface and is covered with luxuriant coral growth is located in the same position and at the same depth as many submerged ridges in the Caribbean area that are barren of coral growth and which Macintyre (1972) has named 'submerged reefs'. Macintyre (1972) feels that these ridges were reefs that formed during the late stages of Holocene sea level rise, but that the reef-building fauna could not keep pace with rising sea level, and so the reef died off. This contention is given strong support by recent drilling and excavation of these ridges off St Croix (Adey, Macintyre & Stuckenrath, 1977) and Florida (Lighty, 1977), both of which are relict reefs of late Holocene age. The only ridge documented to date outside Belize that contains a thriving reef-building fauna occurs off the west coast of Barbados (Macintyre, 1967).

The recurrence of these steps or terraces on the upper part of the profile from many localities indicates that Belize is not unique.

The wall

Because the wall is below the depth of normal SCUBA diving it has, outside Belize, been studied in detail only in Jamaica and the Tongue of the Ocean, Bahamas (Fig. 7-1). In other areas the only information comes from detailed bottom soundings, photographs taken by remote cameras, and systematic dredgings.

The seaward slopes of Funafuti Atoll in the Pacific were the first to be sampled (David *et al.,* 1904). The results indicated that the margin of the atoll is a submarine cliff below which an accumulation of sediment and debris extends seaward. The submarine slope steepens abruptly at a depth of about 82 m from 25 to 30° to 50, 60 and 70°, or in places vertical. From dredged samples it was inferred that the steep slope is encrusted with coralline algae, gorgonians, alcyonarians and molluscs. None the less the surface must be fairly smooth: '. . . as is proved by the fact we repeatedly dragged our dredges over this part of the slope without becoming foul, and we were able to drag the big iron crown with its circle of teeth (each about 4 inches long) over the whole surface without it coming tangled' (David *et al.*, 1904, p. 154).

More recently, Emery *et al.* (1954) described the slopes around the atolls of Bikini, Eniwetok, Rongelap and Rongerik in the Marshalls Group in the Pacific. In general, profiles constructed from echograms and line soundings indicate that the shallow reef extends seaward as a series of spurs and grooves to a gently dipping terrace whose outer edge is at a depth averaging 15 m. The leading edge of this terrace is an abrupt change in slope and the bottom drops from about 15 to 400 m and

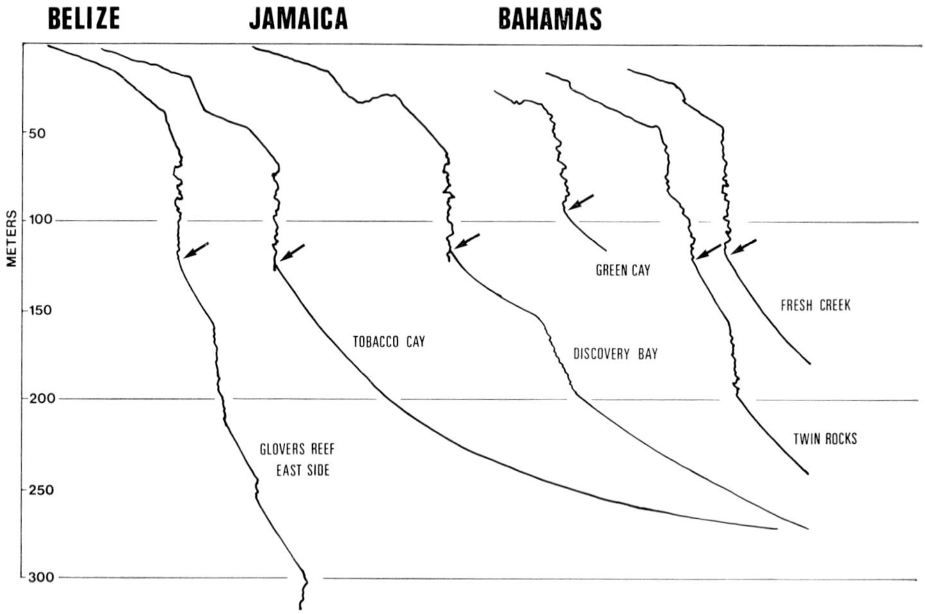

Fig. 7-1. Representative profiles across the platform or reef margin at Belize, Jamaica and Tongue of the Ocean, Bahamas, constructed from submersible observations and depth soundings. The arrow denotes the contact between the wall above and fore-reef below and is always at a depth of *ca* 100–120 m. An irregular cliff-like feature between *ca* 150 and 220 m also recurs in profiles from each area. (Jamaica—Land & Moore, 1977; Bahamas—Schlager *et al.*, in preparation; Belize—this study).

often beyond at an *average* slope of 45°, often steeper. At greater depths, the slope gradually becomes more gentle. This upper steep portion of the reef margin is precipitous along the leeward side of the atolls and in many places can be seen to be a submarine cliff that has a rough, lumpy surface, with encrusting coralline algae, *Halimeda* growth and many ledges veneered with sediment (Emery *et al.*, 1954).

Some of the first direct observations of the reef wall were made by Busby (1962, 1966) and Gibson & Schlee (1967) in the Tongue of the Ocean, a large, deep salient in the Bahama Banks; later Neumann & Ball (1970) described the margins of the Straits of Florida.

The reef wall 'Rim escarpment' of Busby (1962) off Andros Island along the western side of the Tongue of the Ocean extends from the break in slope at 30 m down to 140 m. The surface is a hard rock wall, thoroughly sculptured with notches, caves, and small terraces ranging from a few to 10 m in size and covered with a thin veneer of sand. The notches, though large, do not appear to be continuous. Below a depth of 140 m the wall becomes smoother and continues to about 200 m where its base is buried by sediment and large blocks of limestone.

More recent and detailed observations in the Tongue of the Ocean using the submersible NEKTON (Schlager *et al.*, in preparation) confirm that the wall is almost identical to that off Belize. The wall extends from 50 m to between 125 and 100 m where it is buried by a slope that dips basinward, first at 45° and then gradually flattening to 35°. There is no difference in the morphology of the wall whether it is on the windward or leeward side of the platform or whether it is capped by shallow reefs or by ooid sand shoals.

The most detailed investigation of the wall and deeper sectors of the reef margin outside the Belize area has been carried out along the north coast of Jamaica, first by deep SCUBA diving (Goreau & Hartman, 1963, 1966; Goreau & Goreau, 1973; Goreau & Land, 1974) and later by submersible (Lang, 1974; Moore, Graham & Land, 1976; Land & Moore, 1977).

The wall (or deep fore-reef as it is called in Jamaica) is a rugged, vertically corrugated, near-vertical to overhanging cliff extending from 55 to about 122 m. The surface of the wall is formed by many projecting ledges with numerous holes, caves and fissures. Carbonate sand, rich in *Halimeda*, veneers all projections, floors many caves and is transported down the face both as sinuous streams and free fall. Most promontories can be traced upward into pinnacle reefs on the brow (fore-reef slope) above and most re-entrants are areas of sediment movement down the wall. The promontories are more heavily colonized by organisms than the re-entrants.

Land & Moore (1977) recognize three morphological zones in the wall: (1) 55–70 m: a scleractinian coral reef, dominated by *Agaricia* spp., *Montastraea cavernosa* and several species of *Halimeda*. (2) 70–105 m: abundant sclerosponges, which they infer make a framework, but which are restricted to vertical surfaces and underhands; several large caves are found at 84 m. (3) 105–122 m: a relatively barren area, dominated by soft demosponges and coralline algae.

From the above descriptions, it is clear that the walls of Jamaica and the Bahamas closely resemble the walls off Belize with only minor differences: (1) the vertical corrugations of the Belize walls are not as pronounced as those in Jamaica, (2) in Belize pinnacle reefs at the brow are rare, (3) sclerosponges are much less abundant in Belize and do not, except possibly at site 5, form a framework.

The fore-reef

The base of the wall was first determined by David *et al.* (1904) off Funafuti. They thought that it lay somewhere between 200 and 250 m because at that depth their dredges frequently became fouled. The material recovered was the same as the cliff face: lime sand, pieces of coral and small blocks of limestone. Sediment collected at a depth of 400 m was a sand composed of *Halimeda* joints, coral fragments and foraminifera.

Using a dredge and fathometer, Emery *et al.* (1954), found that the steep slopes off Bikini and nearby Pacific atolls, to a depth of 150 m, are partly covered with large blocks of limestone. Most of the blocks are large fragments of coral, others are composed of *Halimeda* debris, pieces of coral, foraminifera and sand cemented together into a solid mass. Several blocks are 'as well cemented as beachrock' and resemble parts of the shallow reef. The size of blocks appears to decrease with depth, with huge blocks rare below 400 m, but blocks are found as deep as 700 m.

Sediments recovered from the margin are very similar in composition no matter what the depth (Emery *et al.*, 1954). Sand from depths of 65–500 m is either coarse sand composed of *Halimeda* plates or fine sand and silt, or a bimodal mixture of both; foraminifera make up the background in most samples; much of the fine-sand fraction is finely broken *Halimeda* plates. These reef-derived sediments grade seaward into *Globigerina* ooze at a depth of about 2000 m.

Descriptions of the slope that extends basinward from the bottom of the wall, as observed from submersibles off Jamaica (island slope; Land & Moore, 1977) and the

Tongue of the Ocean (Schlager *et al.*, in preparation), are remarkably similar. At the base of the wall in Jamaica, especially in re-entrants, there are piles of coral and blocks of rock perched at angles up to 45°. Little of this coarse debris is seen more than 20 m away from the base of the wall. Below the talus about 40% of the slope in Discovery Bay Canyon, Jamaica is a densely lithified rock with a subdued ridge and swale topography running downslope and overlain in places by a thin layer of sediment that is dammed behind fragments of coral and lithified rock ridges trending across the slope. Many clasts have a diverse population of attached organisms, suggesting relatively infrequent disturbance. The loose sand on the slope is composed of *Halimeda* (particularly species growing on the wall above) with accessory foraminifer, mollusc and coral grains. The slope gradually decreases from 45° at the top to *c.* 25° at 300 m and extends, at an average slope of 10° to a depth of 5200 m where it is buried by sediments of the abyssal plain. There is no recognizable coarse debris below 215 m and the sediments change abruptly at this depth from grey-white reef-derived sediments to red-brown *Globigerina*-rich ooze, with an abundant infauna as evidenced by numerous tracks, trails and burrows (Moore *et al.*, 1976).

This description of the fore-reef (or island slope) could just as easily serve for the Bahamas (Schlager *et al.*, in preparation) with the hard-rock surface composed of innumerable seaward-dipping small shingles dusted with sediment. In both places, a large outcrop or cliff nearly 50 m high occurs between *c.* 150 and 200 m, on the north side of Discovery Bay Canyon, Jamaica and on the south-facing bank margin of the southern Tongue of the Ocean. There are also steep rock walls on the lee side of Glovers Reef (site 5) at this depth (150–180 m).

In the Bahamas the fore-reef slope grades downward into a 'gullied slope' that is a series of spurs and gullies some 200–400 m apart and with pronounced relief running downslope to the flat floor of the Tongue of the Ocean at depths of 1400 metres and more (Schlager *et al.*, 1976). On the upper part of the slope these gullies are canyons tens of metres deep while on the lower parts they are swales with gently sloping banks. This topography is reminiscent of the lower part of the fore-reef off Queen Cays (site 4) and parts of Glovers Reef fore-reef (site 7).

Along the Straits of Florida the slope grades into a smooth rock surface, veneered with moving sand (Neumann & Ball, 1970).

On the whole the fore-reef slopes off the Bahamas and Jamaica are most like those off the leeward side of Glovers Reef (sites 5 and 6). The slope off the barrier reef exhibits no bare rock surfaces and is instead a sediment-covered slope that is much more gentle; bordering the Cayman Trough the slope is steeper with segments like the Bahamas and Jamaica alternating with steep cliffs and sediment-covered slopes. Nonetheless, in all areas the slopes are more alike than they are different, specifically: (1) the break in slope at the base of the wall is always at the same depth, *c.* 110–125 m; (2) limestone blocks and reef debris do not extend very far seaward from the base of the wall; (3) the contact between sediments from the wall and basinal sediments is at about 200 m; (4) the composition of the sediments is about the same; (5) the gentle ridge and swale topography is the same.

The general picture that emerges from all four areas is quite similar to the scarps and talus slopes of mountains on land. The wall is some 60 m high and footed by a relatively narrow, but steep scree slope that flattens rapidly to merge with the horizontal basin floor. What is different about the wall as compared with a terrestrial cliff and its scree slope are the localized accumulations of perched talus and the veil of

sediment, mud and plates of *Halimeda* that covers all subhorizontal surfaces. The piles of coral rubble that are perched here and there in the narrow valleys running down some walls act as dams for finer-grained sediment moving downslope and the large interstices of this rubble trap sediment as well. It is inferred that the rapid cementation of this mix of perched talus and sediment makes a significant contribution to the accretion of the wall, and at the same time the cementation promotes jointing and fissuring that allow calving of blocks of limestone from the wall which accumulate on the slope below.

DISTRIBUTION OF REEF-BUILDING CORALS AND ALGAE

Introduction

Prior to the use of submersibles to study the deeper reaches of reef margins, what was known of the depth zonation and growth forms of hermatypic corals came largely from dredging supplemented by the observations of SCUBA divers in a few areas. Observations from submersibles in Belize as reported here and from Jamaica (Lang, 1974) provide substantially new and more precise information on depth zonation and growth form of both the corals and the principal sediment-producing alga *Halimeda*. In the following sections we summarize our results from Belize and compare them with the observations in Jamaica, Yucatan and the Bahamas as well as earlier obsevations in the Pacific and Indian Oceans.

The growth form of hermatypic corals

A comparison of the variation in growth form with depth for both Atlantic–Caribbean, Pacific and Indian Ocean reefs suggests a recurring trend; branching growth forms predominate in depths shallower than 10 m, massive, domal and columnar growth forms are most prevalent between 10 and 20 m, foliose and platey corals are commonest in deeper water.

Belize

A compilation of the corals observed during dives at the most studied site, Tobacco Cay along the Belize barrier reef, is summarized in Table 7-1. This summary is not meant as a complete list of the corals and their depth ranges, but rather is a summary of many observations and illustrates the general trends of coral distribution. There is a marked change in growth habit from rounded, columnar and massive forms at the top of the step at a depth of about 20 m. Below this depth the massive species such as *Montastraea annularis, Montastraea cavernosa,* and *Porites astreoides* become flatter and more sheet-like. At this depth sheet-like colonies of *Agaricia* species also make their first appearance. Towards the major break in slope at the top of the wall the most apparent corals are plate-like colonies of *Montastraea cavernosa, Mycetophyllia reesii, Agaricia grahamae,* and *Agaricia fragilis,* often growing in colonies over a metre across.

On the wall proper the corals grow as encrusting sheets on ledges with their leading edges often projecting out into space. The three observed on the upper part of the wall are *Stephanocoenia michelini, Agaricia fragilis,* and small branching *Madracis mirabilis.*

Table 7-1. Distribution of different coral species on the reef front off Tobacco Cay (Site 1)

	Spur and Groove		Step		Sand Slope	Brow			Wall
	Middle 15 m	Lower 18–21 m	Upper 20–26 m	Lower 26–35 m	35–45 m	Upper 45–52 m	Middle 52–60 m	Lower 60–70 m	Below 70 m
Astrocoeniidae									
Stephanocenia michelini			X		X			X	X
Pocilloporidae									
Madracis spp.			X		X		X		
Acroporidae									
Acropora cervicornis	XBX								
Agariciidae									
Agaricia agaricites		X	X	X		X	X		
Agaricia lamarcki			X		X				
Agaricia fragilis				X	X		X		X
Agaricia grahamae						X	X	XPX	
Agaricia? undata							X	XPX	X
Siderastreidae									
Siderastrea siderea	X		X	X			X		
Poritidae									
Porites astreoides		X	X	X		X	X	X	
Faviidae									
Dipioria labyrinthiformis	X								
Diploria stigosa	XMX	X	X			X			
Colpophyilia natans	X					X			
Manicina areolata				X					
Montastrea annularis	X	XM&CX	XM&PX		X	X			
Montastrea cavernosa		XM&CX	X	X	X	X	X	XPX	
Meandrinidae									
Meandrina meandrites		X	X						
Dichocoenia stokesii					X				
Mussidae									
Mussa angulosa	X								
Scolymia cubensis					X		X	X	
Mycetophyllia lamarckiana		X	X	X					
Mycetophyllia ferox		X							
Mycetophyllia aliciae			X			X			
Mycetophyllia reesii						X	X	XPX	

X, Coral recorded by observers in submersible and by SCUBA diving.
XXX, The particular coral is a major component of the fauna (B, branching; M, massive; C, columnar; P, plate-like).
——, Depth range of coral species from this data.

Pacific atolls

The only comprehensive data on depth distribution of corals in the Pacific is from dredge hauls.* On the basis of these hauls, Wells (1954) has distinguished three succeeding zones down the seaward slope from the surface algal ridge to a depth of 146 m. The uppermost zone of spurs and grooves extends to a depth of about 18 m and coral growth is prolific only on the lower parts of these spurs, below wave base.

The second zone (Fig. 7-2), beginning at a depth of about 18 m and extending to 91 m, is characterized by abundant foliate growths of *Echinophyllia* and its close relatives *Oxypora* and *Mycedium* (Wells, 1954, 1957). The form *Echinophyllia*, with its relatively thin, plate-like colonies, superficially resembles the flattened forms of *Agaricia*. This zone is transitional between the very shallow reef corals and the strictly deep water forms, and varies in depth from place to place, e.g. 45–75 m off the Maldives (Gardiner, 1903) and Hawaii (Vaughan, 1907). The lower reaches of the zone, at 75–91 m, is the lower limit of active growth.

Recent direct observations from Enewetak indicate that *Acropora* spp. occur to a depth of 50 m because of water clarity. Below 15 m, however, these species assume a tabular shape. In addition, *Pachyseris speciosa* is platey below 30 m (J. Lang, personal communication). These observations suggest that even though the species range varies greatly, the change in growth form is consistent.

Hermatypic corals do grow below 91 m, but their growth is marginal to a maximum depth of about 146 m. The most common coral dredged from these depths is the delicate vasiform species *Leptoseris*. Rosen (1971), in summarizing and updating the distribution of corals outlined by Wells (1954), concluded that coral growth in this deepest, *Leptoseris* zone is even more attenuated in the Indian Ocean than in the Pacific.

Jamaica

Coral growth on deeper parts of the reef margin in the Atlantic is best documented from the north coast of Jamaica (Fig. 7-2). Below a depth of about 25 m on the steep 45–70° slope, shallow-water coral genera change from massive hemispherical forms to flattened colonies, while foliose and branching shallow water genera become spindly and much thinner (Goreau & Hartman, 1963), very likely as a response to attenuated light (Goreau, 1963). These growth forms recall the delicate foliose corals growing in the *Echinophyllia* zone of comparable depth (18–91 m) on the margins of Indo-Pacific reefs. The most abundant species above 45 m off Jamaica is *Montastraea annularis* in flattened growth forms, along with flattened inverted cone-shaped heads of *Montastraea cavernosa* and several species of *Agaricia* (Goreau & Goreau, 1973; Goreau & Land, 1974). The near-vertical to overhanging wall below about 60 m is encrusted by flattened colonies of hermatypical corals, dominated by several species of *Agaricia* (Goreau & Hartman, 1963) with *M. cavernosa, Mycetophyllia reesii,* and *Madracis* sp. occurring as well. The lower depth of coral growth as framebuilders is 75 m, although stragglers have been observed alive to a depth of 98 m (Lang, 1974).

Yucatan

Logan (1969) also recognizes three coral communities with very different growth

* J. Lang has called our attention to the limitations of dredge hauls and the fact that these results will be revised as the result of continuing investigations using SCUBA.

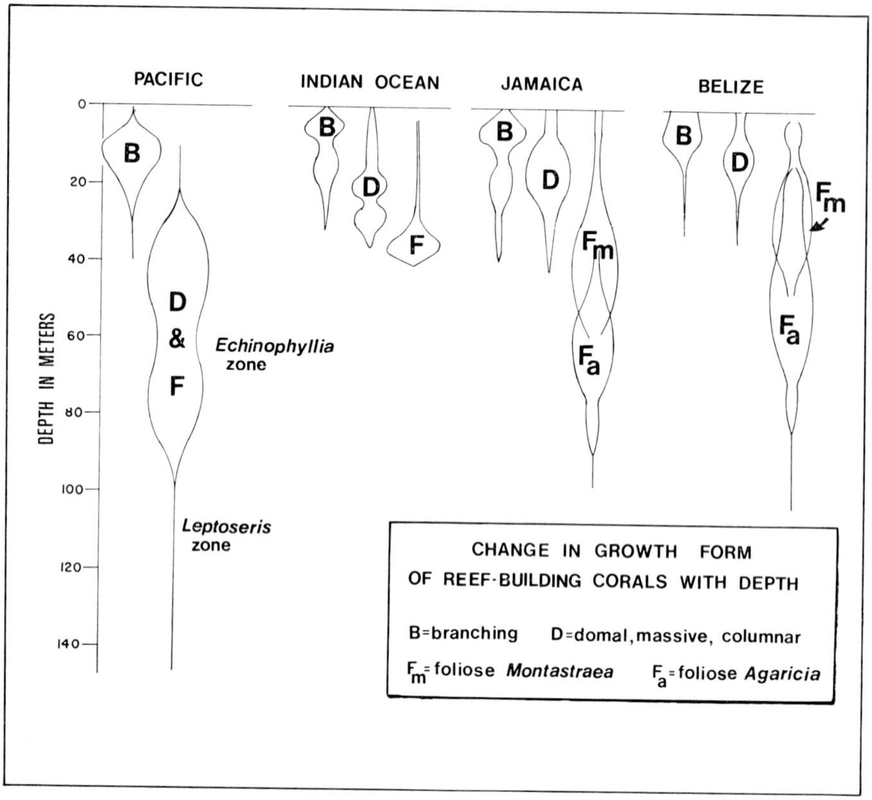

Fig. 7-2. A sketch comparing the depth ranges of the more common growth forms of reef-building corals from four different areas; Pacific Ocean (Wells, 1954, 1957), Indian Ocean–Aldabra (Rosen, 1971), Jamaica (Goreau & Goreau, 1973; Goreau & Land, 1974; Lang, 1974) Belize (James *et al.*, 1976, this study). Corals in the Atlantic–Caribbean grouping are: branching: *Porites porites, Acropora* spp.; massive: *Montastraea annularis, Diploria* spp., *Siderastrea* spp., *Montastraea cavernosa*; foliose *Montastraea cavernosa, Montastraea annularis, Mycetophyllia* spp. and *Agaricia* spp. The diagram is interpretative and constructed from published data only.

forms on the isolated reefs that have formed on the outer parts of the Yucatan shelf: an *Acropora palmata* community of robust branching forms (depth 0–10 m), a *Diploria–Montastraea–Porites* community of domal and hemispherical colonies (depth 5–27 m) and an *Agaricia–Montastraea* community of flattened, plate-like colonies (depth 27–40 m, the floor of the shelf).

Bahamas

Busby (1966) has observed prolific coral growth to a depth of 62 m along the western reef margin of Tongue of the Ocean in the Bahamas. Beyond this depth the amount of coral growth decreases markedly and the deepest living *Agaricia* was recorded at 103 m.

Depth limits of hermatypic corals

The absolute depth of hermatypic coral growth is also illustrated in Fig. 7-2. In the estimation of most observers, hermatypic corals cease to be a significant part of the sessile benthos and can no longer be called reef-building at a depth of between 60 and 80 m. The average depth of 75 m from Belize is surprisingly similar to that off the north coast of Jamaica (Lang, 1974), but shallower than dredge hauls suggest is the deepest limit of reef-building off Pacific atolls (Wells, 1954).

Hermatypic corals grow commonly, though sporadically, to a depth of about 85 m and the deepest living corals were recorded off the eastern side of Glovers Reef at a depth of 103 m, again similar to Jamaica but much shallower than the depth 146 m suggested from dredge hauls in the Pacific.

Halimeda

The distribution of the green alga *Halimeda*, which is responsible for a major part of the sediment on the reef margin, is very similar to that of hermatypic corals. Dredgings around the margin of Funafuti in the Pacific indicate that the alga grows in abundance to depths of 82 m where it decreases in abundance abruptly, and the

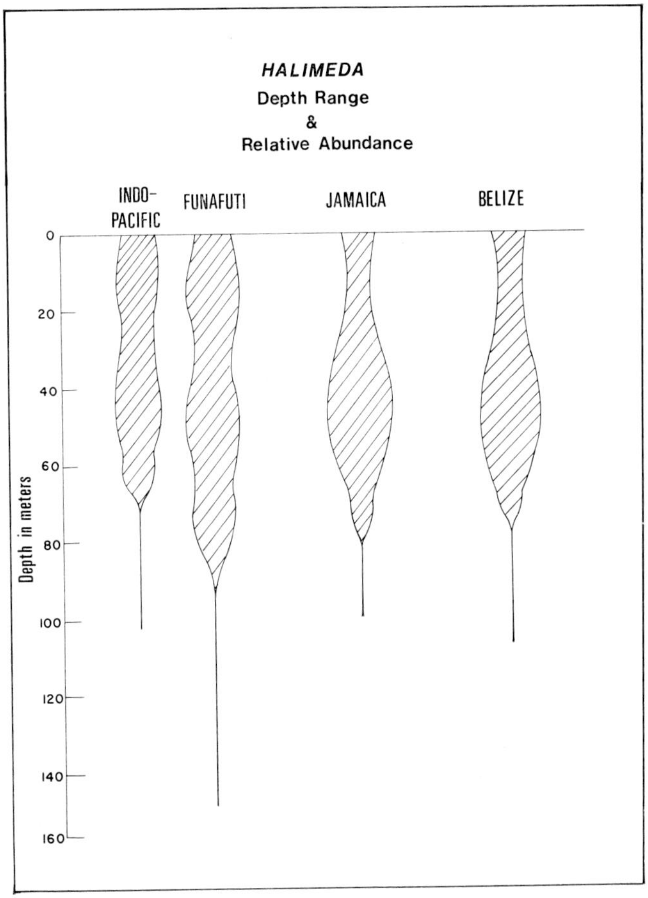

Fig. 7-3. The depth range and relative abundance of the green alga *Halimeda* in four different areas (data from David *et al.*, 1904; Gardiner, 1931; Goreau & Land, 1974; Moore *et al.*, 1976, and this study).

deepest recorded living specimen was recovered from a depth of 146 m (David *et al.*, 1904, p. 135; Fig. 7-3). In the Indo-Pacific area in general *Halimeda* appears to grow profusely to depths of about 73 m, with living representatives found to depths of 110 m (Gardiner, 1931, pp. 77, 79).

Halimeda is extremely abundant on the deeper reaches of the reef margin off Jamaica with growths of *H. copiosa* (Goreau & Graham, 1967) particularly abundant, along with growths of *H. cryptica, H. goreauii, H. discoidea* (Goreau & Land, 1974). These algae grow in profusion to depths of about 70 m where they decrease relatively abruptly although they are still locally abundant (Moore *et al.*, 1976), and only *H. cryptica* extends deeper to a maximum depth of 100 m, Fig. 7-3.

In the Belize complex *Halimeda* is abundant both on sediment slopes and on rock surfaces around and between coral plates throughout the deeper parts of the reef margin above the 65 m deep break in slope. The depth limits of *Halimeda* growth are illustrated in Fig. 7-3. The algae grow prolifically to a depth of between 67 and 73 m, an average of about 70 m. Living algae, however, were seen commonly growing as deep as 80 m and one was observed alive at a depth of 110 m off Glovers Reef.

Reef growth

The distribution and depth limits of corals and green algae, however, do not illustrate an accurate picture of reef growth on the upper part of the margin. Coral growth in particular is not continuous but is limited to linear ridges running down-slope and more commonly to individual mounds, usually metre-size but occasionally up to 5 m high and 10 m across on the steep slopes. Some of the larger mounds are possibly blocks of limestone capped with coral growth. Most of the mounds, however, are located at breaks in slope: (1) at the top of the step at about 20 m; (2) on the abrupt change in slope at the top of the brow at 42–45 m; and (3) at the top of the wall along the barrier reef at *c.* 67 m. This reef growth has a tendency to raise the top of the break in slope slightly, forming a discontinuous rim of coral highs with streams of sand between them. We infer that this coral growth is not simply a veneer, but in places growths that develop preferentially at breaks in slope, not only in shallow water but to depths of 65 m, much deeper than previously suspected.

WALL AND FORE-REEF LIMESTONES: AGE AND ENVIRONMENT OF FORMATION

Age and composition

The earliest report of radiocarbon ages of wall limestones comes from the Grenadines, where D'Anglejan & Mountjoy (1973) dredged limestones from the wall; they reported layered calcarenites and platey corals surrounded by cemented calcarenites with bulk radiocarbon ages of 9750 to 12 900 years B.P.

Following our initial results from Belize (Ginsburg & James, 1973) the same methods were used to sample walls off the north coast of Jamaica (Land & Moore, 1977) and along the margins of the Great Bahama Bank in the Tongue of the Ocean (Schlager *et al.*, in preparation). Analyses of these samples, as well as the ones from Belize, have yielded remarkably similar results both in terms of limestone composition and radiocarbon age (Table 7-2).

Table 7-2. ^{14}C Age of limestones recovered from the deep reef margin in the Atlantic–Caribbean area

Locality*	Reef front	Wall	Fore-reef
Grenadines	—	9750–12 900 (2)	—
Belize	4900–5365 (2)	(a) 2200–2400 (2) (b) 7800–15 220 (11) (c) 7095–12 740 (3) (d) 23 800–32 200 (2)	7856–11 770 (2)
Jamaica	less than 2500 (2)	8600 (1)	9500–12 500 (2)
Bahamas	—	4500–8000 (2)	10 000 (1)
Florida	7595–8900 (3)	—	—
St Croix	6945–9075 (2)	—	—

*For sources see text; (1) the number of analyses; Belize: (a) exterior of wall, (b) wall interior, (c) talus block on fore-reef slope, (d) Glovers Reef locality.

Fore-reef

The hard rock surfaces that form the fore-reef off Jamaica and the Bahamas as well as the similar ridge and furrow structure off Glovers Reef all have radiocarbon ages ranging between 7800 and 12 500 years B.P. In Jamaica this rock is *Halimeda*-rich mudstone well cemented by Mg-calcite. In the Bahamas the composition is similar but the rock contains numerous shallow-water corals. In Belize, although the rock is also composed of massive corals and *Halimeda* packstones to grainstones, the limestone has two aspects which set it apart from the majority of wall limestones; (1) the relative proportion of plate-like corals with sediment sandwiched between them is much higher, and (2) specimens of *Acropora palmata* which never occur in wall limestones, and numerous, large *Acropora cervicornis* sticks which are rare and present only as small, delicate sticks in wall limestones, are common.

Wall

Of the twenty dated samples from the Belize wall, all but four fall in the range 7000–15 200 years B.P., and fifteen of the sixteen have ages between 7000 and 13 000 years B.P. (Table 7-2). Of the two samples that have ages *c.* 2300 years B.P., one is a coral from the outer metre of the wall and the other is the youngest generation of sediment in a cavity filling.

The two limestones with ages older than 15 200 years B.P. come from the base of the wall off Glovers Reef. These rocks, indistinguishable in composition and mineralogy from other wall limestones, have apparent radiocarbon ages of 23 000 and 32 000 years B.P. These ages are near the detection limit of radiocarbon dating and so the samples may be considerably older. Whatever the exact age they clearly formed during a time much earlier than most of the wall limestone and when sea level was lower than today.

All the samples of Belize wall limestone in Table 7-2 have a similar composition: shallow water coral colonies (particularly *Montastraea annularis*) and well cemented, *Halimeda*-rich mudstone to packstone.

The lithified sediments in these wall limestones have radiocarbon ages in the same range as the surrounding corals, an indication that deposition and cementation occurred about the same time as the corals grew, or not long afterwards. Apparently, most of the accretion to the walls in Belize occurred between 7000 and 13 000 years B.P. and since 7000 years B.P. the rate of accretion has decreased significantly.

Reef front

This zone, ranging from the surf zone to the top of the wall is complex and in most places covered with living coral. Excavating and drilling to depths of 2 m subsurface in water 25 m deep off Jamaica (Land & Goreau, 1972) have recovered material with ^{14}C ages up to 2510 years B.P. Excavations of offshore ridges off southern Florida (Lighty, 1977) have revealed a complex series of facies with rocks ranging in age from 7595 to 8900 years B.P. at a depth of 20–30 m. Drilling of an outer shelf-edge ridge off St Croix, which rises to within 20–40 m of the water surface, has revealed not only a Holocene reef, but a vertical succession (Adey *et al.*, 1977). From the top of the ridge to a depth of 3 m subsurface the reef rock is characterized by deep water corals and has a radiocarbon age of 6045 years B.P. Below this zone from 3 to 6 m subsurface is a surf-zone *Acropora palmata* accumulation with a radiocarbon age of 9075 years B.P.

In summary, the outer metres of the wall in all areas studied to date are clearly not erosional, but are an accretionary feature of late Pleistocene to Holocene age. The age dates from all areas indicate that limestone formation and therefore margin accretion can be bracketed in two time periods. The fore-reef limestones and the wall limestones appear to have formed concurrently, between 7000 and 15 000 years B.P., with some minor accretion and erosion continuing until the present day. The thickness of the wall limestone is at least 5 m in Belize and, if the blocks on the fore-reef slope are any measure, may be up to 15 m thick. Limestones at the top of the wall formed more recently, from 9000 years B.P. until the present.

Environment of limestone formation

Evidence from coral fauna

As discussed earlier in this chapter (p. 161) there are marked changes in the growth form, species abundance and species diversity of living corals with increasing depth on the reef front. Using this information and comparing it with the types of corals recovered from the wall limestones we can infer the environments in which limestones formed.

At the one shallow water site, 40 m deep off Glovers Reef, the outer metre of reef rock is a well lithified mixture of plate-like *Agaricia* spp. corals and *Halimeda* packstones to grainstones. This limestone changes inward for at least 5 m to an accumulation of large *Montastraea annularis* colonies, surrounded by unlithified *Halimeda*-rich mud and sand. Although both of these corals grow today at these depths on the surface surrounding the blast site, the *Agaricia* is more common and *Montastraea annularis* is more abundant in slightly shallower water.

The limestone that makes up the outer 4–5 m of the wall proper below 60 m is characterized at all localities in Belize by abundant colonies of *Montastraea annularis* surrounded by *Halimeda*-rich grainstones to mudstones. Many of the corals are in growth position, while others are clearly overturned. The out of place coral colonies may be of local origin or they may have fallen from above. Along with *M. annularis* are other massive to laminar colonies of *M. cavernosa*, *P. astreoides*, *S. radians* and *C. natans*. The shallow-water robust, branching *A. palmata* is absent. The delicate, deep-water *Agaricia* spp. are common only in samples from the outer metre or so of the wall or at the break in slope at the top of the wall.

The absence of *A. palmata* and paucity of *Agaricia* spp. suggest that most of the wall accretion took place in an environment below the shallow surf zone, yet in water shallower than the zones of platey coral growth. The abundance of massive *M. annularis* colonies indicates an environment of intermediate depth, between 10 and 40 m.

Limestones that make up the fore-reef are mainly well lithified sediments, but in the Bahamas and Belize the rocks also contain a wide variety of corals: shallow-water branching species such as *A. palmata* and the slightly deeper water *A. cervicornis*; massive, domal and columnar *M. annularis* and other forms common in wall limestone; foliose to plate-like deep-water corals such as *Agaricia* spp. The occurrence of all these corals together, from different parts of the reef, from shallow water to deep water, indicates that the deposit is, in part at least, allochthonous.

Evidence from sea level curves

There is much uncertainty as to the exact course of sea level rise since the last eustatic lowering during the Classical Wisconsin glaciation (Morner, 1976). We have summarized the published data (Fig. 7-4) using the sea level curves constructed by Fairbridge (1960), Curray (1965), Milliman & Emery (1968), and Morner (1971) and for the very shallow part of the curve, a total of fifteen curves assembled by Curray & Shepard (1972). We have done this because there is no eustatic curve for the Belize

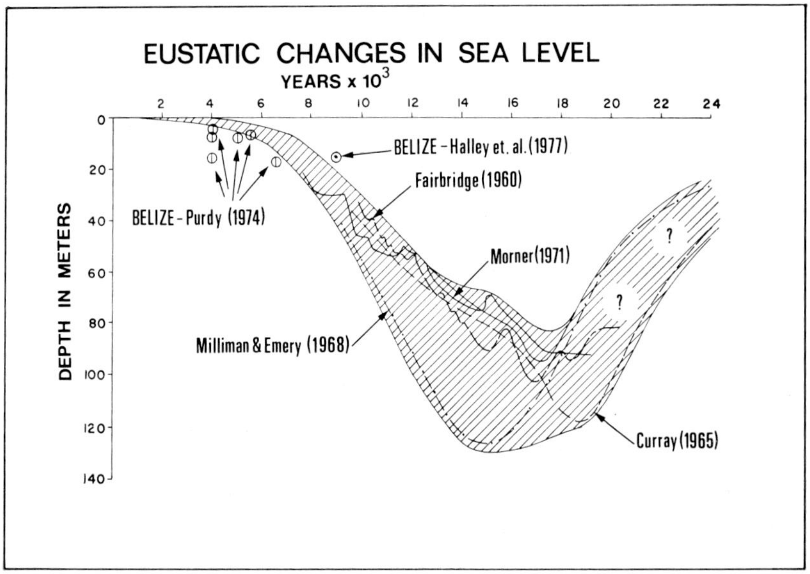

Fig. 7-4. A diagram illustrating the variation in the position of sea level over the last 25 000 years. The envelope that encloses the curve between 8000 years B.P. and present is a summary of fifteen curves from Curray & Shepard (1972) and Neumann (personal communication). Dates older than 18 000 years B.P. are less reliable.

area, fully realizing the dangers implicit in mixing data from different areas. The envelope enclosing all the different curves gives a first-order idea of sea level fluctuations in the last 20 000 years.

Almost all workers agree that the lowest point of sea level during the glacial maximum was *c.* 100–120 m below present sea level, a level roughly coincident with the base of the wall in all areas. This sea level low stand occurred, depending upon the author, between 19 000 and 15 000 years B.P. Although there is disagreement on the fine details, all sea level curves indicate that the rise in sea level took place in two stages: (1) a relatively rapid rise of 100 m between 16 000 and 7000 years B.P. and (2) a relatively slow rise of 20 m between 7000 and the present.

The record of sea level rise since *c.* 7000 years B.P. is of less concern to this study and it also varies markedly from place to place, likely as a result of local fluctuations in the geoid surface (Walcott, 1972; Morner, 1976). On the same diagram (Fig. 7-4) we have plotted dated samples of marine deposits from Belize (Purdy, 1974a; Halley *et al.,* 1977); although the depth of these samples at the time of their formation is not known with certainty and therefore they cannot be used to indicate the position of the sea levels, the dates do indicate that the wall was flooded entirely some 6000 years B.P.

Based on this data it is clear that those corals which are most abundant in the wall limestones grew during a time when sea level was rising from the maximum low stand to a depth about 20 m below that of modern sea level; rising from the junction between the wall and fore-reef, up the wall and over the top. Most of the corals that are found in the fore-reef limestone also grew at this time.

HISTORY OF REEF MARGIN DEVELOPMENT

Based on the preceding information we can now interpret the style of reef-margin accretion (Fig. 7-5).

At the time of maximum glacial advance during the Classical Wisconsin or Wurm II stage, sea level stood some 100–120 m below sea level today. Then the sea would have been washing against the base of a steep, possibly irregular, limestone cliff. If conditions were similar to those today a living reef would extend 60–70 m below that palaeo-sea level, or to a point on the margin some 200 m below present sea level. This situation is similar to what we see today along the north coast of Jamaica and Barbados, where modern reefs grow at the foot of steep limestone cliffs.

We have no direct evidence of a late Pleistocene reef at the base of the wall, only the logical extension of our present data. On seismic records, however, the upper part of the sloping fore-reef all along the barrier reef is characteristically non-reflective and seismically 'transparent'. The seismic signature of reefs in the subsurface is, coincidentally, a seismically transparent zone surrounded by layered reflectors. This area at the foot of the wall then may be either massive unbedded fore-reef debris or a buried reef, or more likely a combination of both, a late Pleistocene reef buried by latest Pleistocene and Holocene talus.

As sea level began to rise, corals and other reef-building organisms became established on the now submerged portions of the 'cliff'. The reef-building community progressively inhabited higher and higher levels on the wall as sea level continued to rise. We have no record of a shallow water, surf-zone community in the wall limestones either because we did not sample far enough back into the wall or because the robust

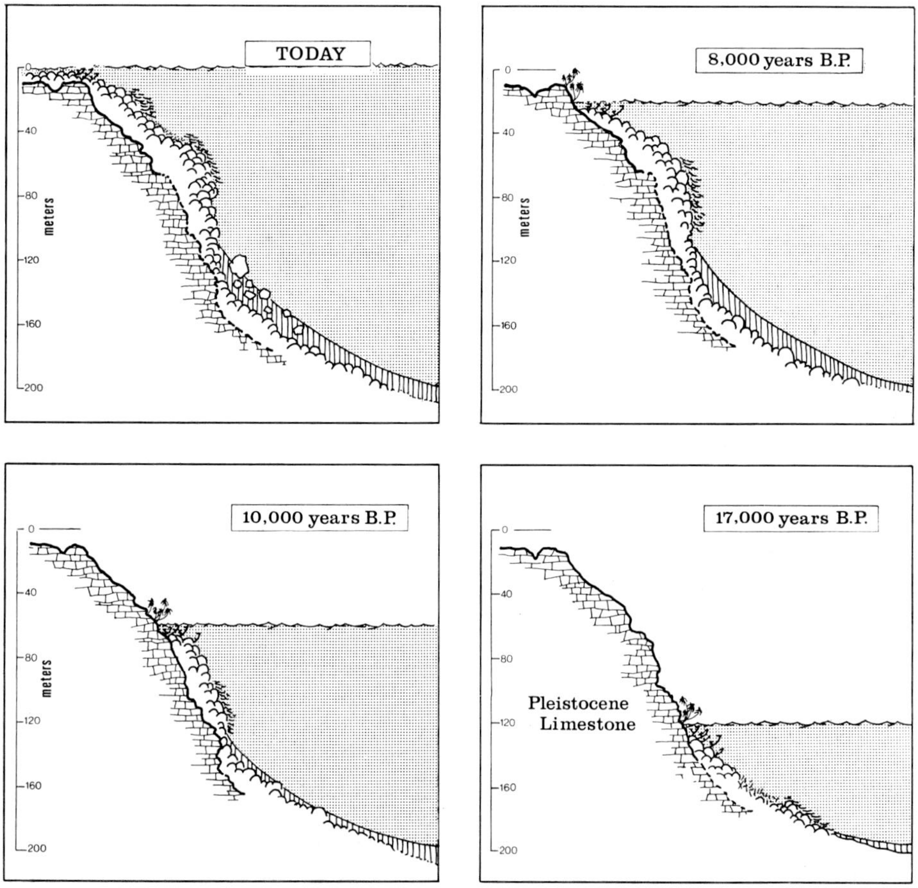

Fig. 7-5. Four sketches illustrating the relative position of sea level and the corresponding zone of reef growth along the Belize barrier reef 17 000, 10 000, 8000 years B.P. and today.

branching species characteristic of this environment were pruned by storm surge and moved down to the fore-reef. The presence of *Acropora palmata* in cemented fore-reef limestones suggests that this last possibility may be the case.

Rather, most of the limestone indicates that maximum accumulation took place in shallow to intermediate depths of water, 10–40 m deep, the zone of maximum growth in modern reefs. There does not, however, appear to be any noticeable change in the age of the limestone as different parts of the wall were sampled. This is because, as the reef grew upward, following the rising sea, material from shallower zones was continually raining down on deeper growing zones, so that any sample is a mixture of corals of slightly different ages. This indicates that the mode of accretion of the wall is compound, the result of (1) the in-place growth of corals and other reef builders, (2) discontinuous accumulations of sediment and debris from above (somewhat later when sea level had risen) that were cemented in place rapidly, and (3) multi-generation internal sediment. Only the samples from the outer parts of the walls show by their ages and by the relatively greater abundance of deep-water platey corals, especially *Agaricia* spp., indications of in-place accumulation in depths greater than 40 m.

With continued rise the sea would have transgressed over the top of the wall (at about 60 m below present sea level), some 10 000–12 000 years B.P.; most of the wall at this time would have been covered with living coral and there would have been a thriving shallow-water reef at the break in slope at the brow with debris from this thriving reef falling and lodging on the wall and on the fore-reef slope below.

By 6000–8000 years B.P. sea level had risen to within *c.* 20 m of present sea level, the top of the Pleistocene limestone ridge that forms the core of the Barrier Reef Platform (Purdy, 1974a). Reef growth at this time would have extended down to the brow and existed on the upper part of the wall to a depth of *c.* 90 m below present sea level. Coral growth on the wall would be restricted to the delicate, slow-growing, deep-water coral species; the time of major wall accretion was over.

Reef growth on the shallower, gently dipping slopes above the wall at this time was complex. In some places reef growth was more or less constant over the whole zone; in other places reef growth was accentuated on or restricted to pre-existing highs. Drilling of some of these reef ridges in other areas has revealed that there is indeed a community succession in shallow water as sea level rose: a shallow water, surf-zone community is replaced by an intermediate water depth community (Adey *et al.*, 1977). These data together with the excavations from Belize suggests that the modern distribution of coral zones was established about 5000–6000 years B.P. Since this time, upward growth has been at least 20 m at the reef crest (Purdy, 1974a).

MORPHOLOGY OF THE MARGIN: A MODEL OF DISCONTINUOUS LATERAL ACCRETION

Because we believe there has been significant accretion to the reef margin during the Holocene rise of the sea level, we infer that similar accretions developed during preceding periods of rising sea level. Each of these accretions was exposed to sub-aerial erosion at low stands of sea level and was certainly eroded. Whether or not each period of erosion removed the preceding accretion or merely sculptured it is unknown, but it is our opinion that accretion predominated.

It is generally believed that sea level last stood near its present position some 80 000–120 000 years ago (Bloom *et al.*, 1974). We assume that the Pleistocene limestone found 20 m below present sea level on the Barrier Platform (Purdy, 1974a) was deposited some 80 000 years ago. In the period between 80 000 and 10 000 years B.P. sea level was not constantly low, *c.* 120 m, but as continental glaciers waxed and waned so sea level fluctuated up and down. The high points on the sea level curve between the major interglacial (Sangamon) *c.* 125 000–80 000 years B.P. and the present interglacial (Fig. 7-6) are between 25 and 40 m below present sea level (Bloom *et al.*, 1974) and correspond to well known interstadials as recorded at the margin of the Laurentide ice sheet (Dreimanis & Karrow, 1972).

During parts of each of these three fluctuations of sea level, both rising and falling, it seems likely that the wall would have accreted seaward. We can only guess how much accretion and how much erosion occurred during each of the three fluctuations preceeding the Holocene. If, however, we use our estimate of 10 m of accretion during the Holocene as a guide, and we accept the inference from Fig. 7-5 that the periods of low stand of sea level were relatively short, we conclude that the wall is built of successive accretions.

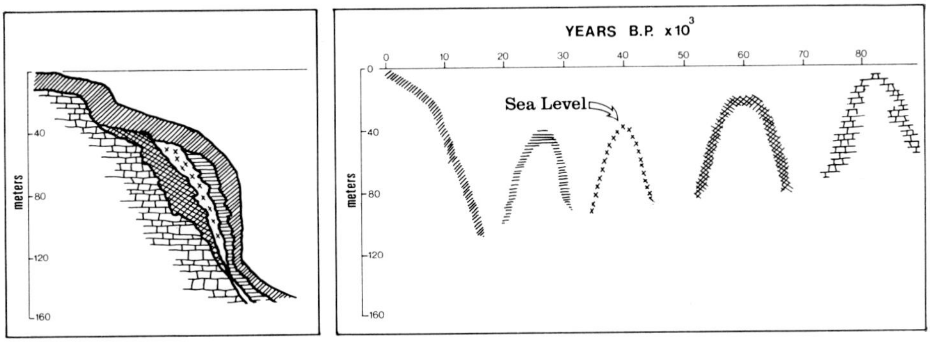

Fig. 7-6. A sketch illustrating the fluctuations of sea level (right) during the last 80 000 years (from Bloom *et al.*, 1974) and the lateral accretion (left) of the margin corresponding to each individual rise and fall of sea level.

The present morphology of the margin is likely a reflection of this style of accretion. The coincidence of the step in morphology at about 30 m and gentle sand slope likely reflects the upper level of reef growth during sea level high stands, and is not wholly an erosional feature. This feature is now subdued by the overgrowing modern reef facies. This feature is world-wide (see page. 156), suggesting that this style of accretion may be a universal phenomenon.

An alternate hypothesis has been proposed by Schlager *et al.* (in preparation) to explain the wall on the basis of observations in the Tongue of the Ocean, Bahamas. They argue that the wall owes its morphology primarily to coastal erosion during low stands of sea level during the Pleistocene. Thus the material sampled on the modern wall is merely a thin veneer over a Pleistocene or older erosional surface. Any pre-existing accretion formed during other Pleistocene rising sea levels was stripped away in a like fashion during the following sea level low stand.

In the final analysis both hypotheses are possible; we favour the accretionary model, but only drilling can supply evidence for either interpretation.

Chapter 8

Sedimentation and diagenesis on the deep seaward margin of modern reefs

INTRODUCTION

In the preceding chapters we have dealt with the deep reef margin in general terms; now we focus on two particular aspects that have direct application to the interpretation of ancient limestones; sedimentation and diagenesis. Using our findings, and comparing them to those of others, we outline the origin and dispersal of sediments on the deep margin, compare the composition and fabric of these modern sediments to the latest Pleistocene and Holocene limestones that make up the wall and compare the types of cements and diagenesis discovered in these rocks with findings from other areas.

ORIGIN OF SEDIMENTS

Sediments on the proximal fore-reef, a melange of limestone blocks and coral debris in a matrix of sand, are all derived from the deep reef and wall. Limestone blocks that come from the wall above (see Chapter 5) have fallen onto the fore-reef and perhaps rolled seaward. The numerous corals are mostly the plates of *Montastraea* spp. and *Agaricia* spp. which grow on the deep reef that crowns the wall and have fallen onto the slope below. These corals with plate-like colony shape often have their small and narrow basal attachments intensively bored and thus weakened by endolithic sponges and bivalves (Goreau & Hartman, 1963) and so are easily detached. Shallow-water corals are not now transported to the fore-reef; no characteristic species such as *Acropora palmata* or *Acropora cervicornis* were seen in any of the dives along the margin nor recovered in any samples, except in very deep water off Glovers Reef, where they are clearly resedimented, older deposits (see page 57–59).

Samples of sediments on and within the wall and fore-reef are composed of three grain populations: (1) granule-size, composed almost entirely of *Halimeda* plates, particularly the species that grow in intermediate to deep water; (2) coarse to fine sand size, composed of broken *Halimeda* plates, coral fragments, mollusc grains, coralline algae, benthic foraminifers and lithoclasts; (3) very fine sand and silt, composed of sponge chips, spicules of various kinds and small clasts.

When these sediments were analysed, one of the surprises was the abundance of silt-sized fragments contributed by the boring of endolithic sponges. Both in Belize (this study) and in Jamaica (Moore *et al.*, 1976) these sponge chips comprise *c.* 1/3 of the silt-sized fraction, and depending upon the grain size of the whole sediment sample, form from 1 to 12% of the sediment. Endolithic sponges are common in all parts of the living reef and on the wall, and have been seen in limestone blocks on the fore-reef slope, so that chips may be derived from all three sources.

The lack of shallow-water coral species in the rubble on the fore-reef slope as well as the lack of shallow-water *Halimeda* fragments suggests that at present little coarse-grained sediment generated on the shallow reef is deposited on the fore-reef slope. In the finer-grained size fractions there are fragments of *Homotrema*, a foraminifer that grows only on hard substrates, especially those associated with reefs (Emiliani, 1951; James *et al.* 1976). *Homotrema* exhibits a variety of growth forms (Emiliani, 1951; Rooney, 1970) and although it grows in deep water is much less common and changes shape from globular to slender and branching in deeper water. Mackenzie *et al.* (1965) have used the presence of *Homotrema* tests to illustrate that shallow-water sediments are transported seawards considerable distances from the shallow Bermuda reefs and Chave *et al.* (1962) have reported *Homotrema* in samples over 50 km from the Bermuda reefs at abyssal depths. *Homotrema* in fore-reef sediments from Belize, although present, is far from common. Most of the grains that are in the sediment are from the branching and slender growth forms, suggesting derivation from the deep reef.

Our conclusion about the origin of the fore-reef sediments is the same as that resulting from analyses of sediments from the fore-reef slope (island slope) of Jamaica, where Moore *et al.* (1976) have concluded that most of the sediment on the deep reef margin comes from the deeper parts of the living reef.

In summary, sediments on the wall and fore-reef are derived from the intermediate to deep-water zones of the living reef, and to a lesser degree from the wall itself. Evidently sediments formed in shallow water, *c.* 10 to possibly 20 m, either remain in place or are transported shelfward. The occurrence of large blocks on the reef flat, sheets of coral rubble in the lee of the barrier reef and vast areas of *Halimeda*-rich sand over much of the barrier platform, shows that shelfward transport is a major process.

DISPERSAL OF SEDIMENTS

The sloping fore-reef

The living reef above the wall is a prolific sediment producer, as evidenced by the cover of sand over almost all but the most elevated parts of the reef. From the submersible we could see that some of this sand was moving downslope, as winding streams between coral mounds and promontories. Where the brow was steep, cascades of sand were often seen spilling over the edge from above.

The sediment usually moves down the wall from ledge to ledge, so that the upper surface of each and every projection is covered with a thick wedge of white sediment. Streams of sand can also be seen moving down re-entrants in the wall. Fine sediment must also trickle down inside the wall as many caves, 3 m or more in from the surface of the wall, are floored with fine sediment. In total, there is a continuous rain of sediment onto the fore-reef slope, punctuated by occasional dislodged corals dropping from above and large rock-falls.

This sediment accumulates on the fore-reef below, but does not move far basinwards. Cones and wedges of talus are piled up against the wall, like a series of overlapping alluvial fans. These cones are highest in the re-entrants between rock bosses, suggesting that there is some funnelling of sediment down these recesses in the wall. The fans and surrounding halo of limestone blocks and coral debris do not extend down and away from the base of the wall for any great distance, rarely deeper than 150 m. No trains of limestone blocks were observed trending downslope. The blocks and upturned coral plates act as sediment dams; sand is always piled up behind them on the upslope side, and sometimes overtops the obstruction, partially burying it. The blocks themselves are heavily encrusted with epiphytes, indicating that they have been stationary for some time.

The streams of sand that course downslope between blocks often extend beyond the zone of blocks onto the deeper sand slopes as small lobes of white sand outlined against the generally grey bottom. Small slump scars are occasionally seen on the sand slope but the sediment does not move very far basinward.

On the gently dipping distal fore-reef, evidence of physical downslope sediment movement is absent and the bottom is dotted with animal burrows and trails as animals bring sediment to the surface; each resulting mound is slightly asymmetric with a gentle downslope side, so there is continuous, slow, animal-induced sediment creep.

This lack of significant downslope movement in this type of fore-reef setting is confirmed by petrographic and grain-size analysis of the sediments; less than 4·2 km away from the wall in depths of 200 m (about 80 m below the level of the base of the wall) the sediments are carbonate muds rich in planktonic foraminifers with no recognizable reef or wall derived particles in the sand or silt size range.

None of our dives was located opposite the passes through the shallow reef so that we could not assess the contention of Purdy *et al.* (1975) that there is considerable movement of shallow-water sediments basinward through the reef passes. Clearly though, opposite the long streches of living reef, there is virtually none.

Sediment recovered in piston cores up to 5 m long on the distal fore-reef and margin of the shallow trough is similar to sediment on the surrounding sea floor at comparable depths except for several layers, up to 50 cm thick, composed of coarse-grained sediment that is sometimes graded. The sediment in these layers is mainly skeletal or intraclastic; the skeletons are generally of deep water origin while the intraclasts are comparable in composition to the sediments on the sea floor, but they are cemented by Mg-calcite micrite. One intraclast has a radiocarbon age of about 16 000 years B.P. These cemented clasts are probably derived from partially cemented hardgrounds or small cliffs related to faulting on the distal fore-reef.

The lack of sediment containing abundant particles of shallow-water origin leads us to conclude that basinward transport of reef-derived sediment is not now a major process or that gravity-induced sediment flows are bypassing the distal fore-reef and coming to rest out in the basin.

In Jamaica, most downslope sediment movement occurs in the axis of Discovery Bay Canyon (Moore *et al.*, 1976). Even though there is a strong gravitational component to sediment movement, because the slope remains steep, up to 24° at 305 m depth, no reef-derived sediment appears to reach the depths of the Cayman Trench. Moore *et al.* (1976) conclude that much of it is trapped behind large blocks or outcrops (called 'Haystacks') on the deeper reaches of the slope. Land (1979) concluded that a great deal is dissolved in corrosive bottom waters.

The cliffed fore-reef

All of the observations on the steep fore-reef off Glovers Reef and the southern part of the barrier reef give the impression of active, episodic and continuing downslope sediment movement. In places there is no fore-reef slope and the wall continues but where the slope is present the cones of talus that bury the wall dip steeply, often over 30°. The ridge and swale rock slopes are only dusted with sediment that is rich in *Halimeda* to the deepest depths observed, 310 m. Coral rubble and blocks are seen all down the slope, perched precariously or dammed up one behind another. At one place we found a broad amphitheatre on the wall in which the 'floor' was a field of metre-size coral boulders, rounded heads of *M. annularis* and *Diploria* sp. with sticks of *A. cervicornis* and blades of *Millepora* sp.

The transport of these sediments into very deep water appears to be occurring off Glovers Reef because the bottom sediments in water depths of 2200 m, 30 km away from the reef, contain significant amounts of shallow-water derived particles. In contrast, southward along the barrier reef seaward transport appears to be minimal because no such particles are recognized in the deep water deposits. Regardless of location, however, these sediments contain progressively decreasing amounts of calcium carbonate with increasing depth, below 1000 m. This may be due to: (1) dilution by terrigenous clastic sediments carried out over the shelf during low stands of sea level in the Pleistocene, (2) trapping of the carbonate high up on the slope, or (3) dissolution of the carbonate by corrosive bottom waters, as Land (1979) suggests for a similar situation off Jamaica.

SUMMARY

Where the fore-reef is gentle and flattens with depth, grading into a relatively shallow (*c.* 500 m deep) basin, we found no evidence of sediment movement into deeper water. The most likely explanation is that the wall traps much of the sediment moving down it from the deep reef and that what sediment does reach the fore-reef slope consists of poorly sorted and angular grains that are not easily mobilized. Another possibility is that the sediment is moved as gravity induced mass movements and bypasses the distal fore-reef en route to the axial part of the basin.

Where the fore-reef is steep and continues so to great depth, such as along the margins of the Cayman Trough off Belize, the profile is oversteepened and sediment accumulations are subject to episodic mass movements, transporting material into very deep water and great distances away from the wall. A similar situation is present along the margins of the Tongue of the Ocean, Bahamas, where shallow water sediments ranging from oolitic sands to huge blocks of wall limestone are observed at the junction of the steep margin and flat floor of the salient in depths of 2000 m and more (James, personal observation). In addition, carbonate turbidites are commonly recorded in cores from the Tongue of the Ocean (Kier & Pilkey, 1971; W. Schlager, personal communication).

SEDIMENTS IN REEF MARGIN LIMESTONES

Partly to wholly lithified sediments that form the matrix between coral colonies in limestones from the wall, talus blocks and cemented parts of the fore-reef are always multigeneration. This fabric is easily seee in most samples where different generations of deposition can be separated by differences in composition, contact relationships and striking differences in hardness. While the well lithified sediments have radiocarbon ages that generally range from 7000 to 15 500 years B.P., analysis of one friable *Halimeda*-rich sand deposited as a last stage cavity filling in a multigeneration limestone gives an age of 2235 years B.P. Because present surface sediments have an age of *c.* 835 years B.P. this date indicates that multigeneration sedimentation in cavities is a continuing process, not restricted to the period between 7000 and 15 500 years B.P.

The origin of these sediments is both 'primary', formed by the whole or spontaneously disaggregated skeletons of segmented calcareous benthos, and 'secondary', fragments of skeletons and limestone, eroded by boring sponges. At first glance these sediments have the same composition as modern surface sediments on the reef margin; many *Halimeda* plates and a matrix of fine-grained sand and mud. The *Halimeda* plates are characteristically from deeper water species *H. cryptica* and *H. copiosa* and the fine-grained sediments are composed of the same elements as the unlithified surface sediments. The mud-sized fraction is again rich in silt with little clay-sized carbonate.

The texture of the limestone results from complete mixing of the three grain populations and is the same as that of surface sediment samples taken from ledges and caves in and on the wall. In this way they are unlike sediments both on the adjacent proximal fore-reef, which are coarse-grained, and on the distal fore-reef, which are fine-grained. This texture is, however, identical to that of the limestones that form the interior parts of the reefs in shallow water (James *et al.*, 1976). The sediments in the two areas differ only in the composition of the coarse-grained fraction; shallow reef samples contain many coarse-grained coral fragments and numerous tests of the foraminifer *Homotrema*; wall samples are dominated by the presence of large *Halimeda* plates.

The multigeneration nature of the sediments, as well as their composition, indicates that sediment is continually filtering down through the reef mass by way of the tortuous interior cavity network that characterizes the limestones. As a result most of the sediments are truly 'internal', on all levels. This internal sediment fabric is characterized by well laminated sediments with geopetal (but not always horizontal) fabric filling various holes. It can be seen in all primary 'growth' cavities, inside and between corals, as well as in secondary cavities, the result of excavation by bioeroders cut into skeletons and cemented limestones. On a finer scale it also exists in the sediment infill itself, as the fine-grained sands and silts filter down and accumulate in the spaces in the 'house-of-cards' structure created by *Halimeda* plates.

Once deposited, the finer-grained sediments were habitats for various infauna as evidenced by the bioturbated nature of the sediments similar to the observations in cavities of shallow Bermuda cup-reefs (Ginsburg & Schroeder, 1973).

In summary, the texture of sediments *within* the reef or wall is everywhere similar, from the surf zone to the base of the wall; the sediments contain abundant fine-grained material, they are deposited internally and they are often multigeneration. Internal sediments from different reef zones, however, can be differentiated on the basis of grain composition. In contrast to the internal sediments, surface sediments surrounding the reef are coarse grained and contain virtually no fine-grained fraction.

CEMENTATION

Carbonate cement in Belize platform margin limestone is either aragonite or Mg-calcite. *All* the individual cement morphologies found have been reported from subsea synsedimentary limestones in other areas, particularly reefs (Macintyre *et al.*, 1968; Land & Goreau, 1970; Ginsberg *et al.*, 1971; James *et al.*, 1976), open shelf sand sheets (Shinn, 1969), ooid shoals (Ball, 1967) and deep sea environments (Milliman, 1974, pp. 288–300; Marlowe, 1971; Gervitz & Friedman, 1966). What set these Belize synsedimentary limestones apart from the other occurrences of subsea cementation, however, are both the extent and degree of lithification and the variety of cements. The complete spectrum of these cements in one place indicates that the variations we see are locally induced and not the result of differences in chemistry, circulation or substrate.

Mg-calcite

Mg-calcite, the most common cement, binds sedimentary grains together and partially or completely fills cavities. There are three principal forms of Mg-calcite cement, which in order of importance as 'rock-making' cement are (1) micrite (2) bladed spar, (3) blocky spar. The distribution of micrite and bladed spar is in general related to pore size, with micrite most common in small pores and bladed spar in larger pores, similar to the distribution in shallow Belize reefs (James *et al.*, 1976). Bladed spar may vary from short 'stubby' almost equant crystals to long fibrous crystals, but the same elongate bladed nature of the basic crystals is present throughout. Blocky spar is rare and unimportant as a cement.

Growth of bladed spar, especially as thick isopachous fringes, has been recorded from most subsea cemented limestones (Macintyre *et al.*, 1968; Land & Goreau, 1970; Marlowe, 1971; Schroeder, 1972; James *et al.*, 1976) and is generally recognized as a fringe of crystals growing normal to the substrate. The thick and widespread nature of the fringes in the Belize limestones has enabled us to recognize that the growth of bladed spar is commonly spherulitic, with numerous small splays, nucleated at specific intervals along the substrate, forming a series of small fans that coalesce with distance away from the substrate to form a continuous fringe.

Aragonite

Most aragonite in Holocene reef limestones is passive cavity-filling cement (Friedman *et al.*, 1971; Schroeder, 1972; James *et al.*, 1976). Aragonite as an interparticle cement is more common in ooid shoals (Ball, 1967), in sand sheets on the floor of the Persian Gulf (Shinn, 1969), in beachrock (Ginsburg, 1953; Stoddart & Cann, 1965) and in cemented layers on the floor of the Red Sea in deep water (Gervitz & Friedman, 1966). In the Belize wall and fore-reef limestones aragonite is not only an important interparticle 'rock-making' cement but it also develops distinctive botryoidal growths.

The two crystal forms of aragonite cement are long needles and small equate crystals that we term blocky spar; only the needle-like crystals are important as a cement, blocky spar is rare. The needles grow in three different ways, producing three different cement styles: (1) a mesh of needles between particles, (2) epitaxial extensions of skeletal aragonite crystals projecting into interparticle voids, and (3) partial to almost complete spherulites that we have termed botryoidal aragonite.

Aragonite mesh cement is unlike the aragonite cements in ooid shoals, which are isopachous fringes with needles growing normal to the grain surface, but is like cements in subtidal sands of the Persian Gulf (Shinn, 1969) and the arrays of aragonite fibres growing out from centres of accretion in hard layers on the bottom of the Red Sea (Friedman, Schneidermann & Gervitz, 1971). In most Holocene reef limestones aragonite is usually precipitated in coral or gastropod cavities (as it is also in these limestones) as syntaxial extensions of skeletal crystallites. Spherulitic aragonite growth has to date been observed on a very limited scale in Persian Gulf calcarenites (Shinn, 1969) and as cavity fillings in Bermuda reefs (Schroeder, 1972).

Sediment grains and/or layers of sediment often occur within botryoidal aragonite. Bands of sediment are parallel to the convexity of the arrays. There are three possible explanations for the presence of these particles: (1) displacive crystallization, with grains being carried up and spread apart by the cement as it grew, (2) replacement of micrite by aragonite needles, or (3) contemporaneous cement growth and sediment infiltration so that grains are deposited on the growing spherulitic cement. It is hard to support the possibility of displacive crystallization on two counts. First, it is difficult to envisage how some grains would be carried upwards as the cement grew while others were left behind to yield a cement botryoid studded with particles. It is also difficult to see how layers of sediment within the cement, which are grain supported, could have been moved upwards and outwards from their original horizontal position and still remain in grain contact. As a result we feel that the most logical explanation is the growth of aragonite spherulites at the same time as sediment rained down from above. A similar fabric can be seen in smaller cavities, where particles 'float' in aragonite needles. At first glance these particles might appear to be remnants of recrystallization of some grains to spar, but instead they are grains that have fallen into cavities where cement was growing.

The grain-size of the sediments and therefore the pore size appears to govern the crystal size of the cement; micrite sized Mg-calcite occurs most commonly in silt-sized sediments, bladed Mg-calcite and all aragonite cements are commonest in the pores of sand-sized sediments. The reasons why some sand-sized sediments are cemented by and some cavities lined by Mg-calcite spar and some by aragonite is unclear, but the intimate inter-relationships of cement layers indicate that the control is a subtle one. These complex cement patterns, partly the result of the complicated grain fabric, result in an extremely irregular and variable distribution of cements. This patchy distribution of aragonite and Mg-calcite cements, often adjacent to one another in sediments of apparently identical composition, resembles a situation in which one of the cement minerals is recrystallizing to the other phase (i.e. Mg-calcite to aragonite) as suggested by Evamy (1973) and Taylor & Illing (1971) from studies of certain shallow-water carbonates. Careful examination of the Belize cement fabrics, however, reveals that in many cases there are still voids between crystals in both cements and that interparticle spaces are often not completely filled with cement. Such relationships should not be present if recrystallization has taken place.

In summary, displacive crystal growth and recrystallization do not appear to have taken place in these limestones. The complex and often puzzling cement–sediment inter-relationships result first from a complex pattern of cementation, partly induced by the wide variation in pore sizes, and second from the growth of cement in cavities at the same time as fine-grained sediments are trickling into the voids.

ALTERATION

The one aspect that distinguishes wall and fore-reef limestones is the ubiquitous early transformation of skeletons and lithified sediments to mudstone. This alteration is brought about by multiple generations of boring, sedimentation and lithification; cavities are eroded into recently cemented sediments or perished coral colonies by a variety of endolithic organisms; sediments, generally fine-grained, filter into the cavities and are lithified quickly. By this process limestones with a depositional fabric that ranges from coral framestone to coral bindstone to skeletal grainstone are changed to mottled packstones to mudstones.

This process of alteration is common in Holocene reefs; it is known from algal cup reefs of the Bermuda platform (Ginsburg & Schroeder, 1973) and from shallow fringing reefs in the Bahamas (Zankl & Schroeder, 1972), but in neither of these locations has alteration developed to the extent seen in the Belize limestones.

Much of the fine-grained sediment that fills cavities is eroded from limestones and corals by endolithic sponges; as soon as these chips are deposited and cemented they can be bored again, giving multi-cycle grains. It is ironic that cementation, at the same time as it tends to preserve skeletal grains and sponge chips by cementing them into rock, sows the seeds of their eventual erosion by creating hard substrates into which endolithic organisms can bore.

Because the limestones are formed and altered in roughly the same environment, the sediments that filter into the excavated cavities are of nearly the same composition as the original rock. Thus many rocks display what appear to be bioturbation structures, yet are really several generations of alteration. One of the characteristic fabrics in these rocks, however, is geopetal cavity fillings of several generations of cemented marine sediment in the lower half of a cavity with a record of each generation reflected on the roof of the cavity by a layer of cement. This fabric is a key texture that has been used so often in ancient reefs (Schmidt, 1971; Zankl, 1971; Krebs, 1974) to recognize synsedimentary lithification.

The presence of iron oxide and manganese oxide coatings in many cavities either on the walls of empty cavities, or separating generations of sediment in cavities, is important to the interpretation of diagenetic sequences.

The clearly oxidizing conditions in which these coatings have formed is consistent with descriptions of other subsea cemented sediments, occurring as buff to white coloured sediment.

The age, lithology, mineralogy and correlation with our knowledge of sea level changes over the last 20 000 years all indicate that these limestones have never been subaerially exposed. There is a striking similarity between the brick-red coatings formed below sea level with *terra rosa* and oxide coatings so common on and in limestones exposed subaerially. Clearly the presence of iron oxide coatings alone in ancient limestones is an undependable criterion for subaerial exposure.

The relative concentration of manganese and iron in sea water is relatively low. It has been estimated that in deep water it would take 1 million years for a 1–4 mm thick manganese crust to form (Ku & Broecker, 1969). In these limestones, most of which are younger than 15 000 years B.P., crusts 0·5 mm thick are common. In addition, some of these coatings must have formed relatively rapidly, because the cavity has subsequently been filled with sediment which is now lithified.

These metal oxide coatings record times when sediments did not filter into the cavities and carbonate did not precipitate on the cavity walls, indicating that sedimentation and carbonate precipitation are not constant processes within the wall and below the surface of the fore-reef slope, but instead are episodic.

DISCUSSION

All of our observational, petrographic and chemical data indicate that cementation of sediments on the deep reef margin takes place early, in the environment of deposition, and is intense, though not necessarily pervasive. These discoveries extend the range of synsedimentary cementation to fill that gap between the style of cementation found in shallow reefs and the type of cementation within hardgrounds of deeper water.

In the preceding chapter we concluded that the large scale morphology of the deep reef margin was due to accretion during the late Pleistocene and Holocene. Most of the smaller scale, second order morphological elements are likely due to the interplay between submarine cementation and bioerosion. Cementation is not pervasive, and so parts of the wall are well cemented and parts of the wall are unlithified. This irregular lithology would lead to washout of the poorly lithified material and so enhancement of the lithified portions, resulting, in part, in the layered appearance of the wall. In addition, cement adds to the density of the often over-steepened parts of the wall and makes the whole structure brittle. Blocks calve off the wall both because of their own weight, especially when overhanging, and because of the occasional seismic shocks and, possibly, storms in the area. The calving off of blocks both creates reef talus and adds to the vertical nature of the wall. Thus cementation is a critical element in the formation of the structure, on the one hand leading to lithification and accretion and on the other leading indirectly to the formation of talus and accentuating the steepness of the wall.

While initially formed by reef growth together with sedimentation and cementation, the wall also is continuously eroded by submarine processes. The overwhelming evidence of biological erosion indicates that much of the small-scale, sharp and irregular morphology is the result of boring by endolithic sponges and to a lesser degree bivalves and worms.

The signature of these wall and fore-reef rocks lies in their somewhat peculiar composition and the rapidity by which they are cemented. Studies of the shallow-water Belize reefs (James *et al.*, 1976) has convinced us that cementation is both facies and fabric controlled, that reefs and particularly reef associated sediments that have a matrix of internal sediment are preferred sites for precipitation of Mg-calcite cement. This study allows us to characterize even more specifically the parameters that distinguish areas in which early cementation is localized.

One of our major findings has been the similarity between the style of cementation in wall and fore-reef limestones and hardgrounds. The two attributes that appear to make hardgrounds preferred sites of cement precipitation are: (1) they are sites of non-deposition and (2) the sediments are stabilized in some way. The internal sediments of wall limestones satisfy both of these prerequisites. Their internal nature precludes movement once deposited, except perhaps by bioturbation. The grading and iron–manganese coatings in cavity fillings indicate that sedimentation is episodic, so there are periods of deposition alternating with periods of non-deposition.

Our samples have confirmed that subsea cementation by aragonite and Mg-calcite occurs to depths of at least 175 m, and off Jamaica Mg-calcite cemented samples have been recovered at a depth of 280 m (Land & Moore, 1977). The lower limit of Mg-calcite precipitation is defined by the occurrence of Mg-calcite cement inside the tests of planktonic microfossils. Above a depth of 600 m the tests are commonly filled with Mg-calcite cement, below 600 m they are empty. We do not have any ^{14}C dates for these sediments but their presence at the surface of the sediment indicates that they are at least late Pleistocene in age. Thus our data suggest that precipitation of cement occurs to a depth of 600 m, but not beyond, and that Mg-calcite and aragonite are important cements to a depth of at least 175 m.

These conclusions agree with the findings of Schlager & James (1978) who have compiled the data available on $MgCO_3$ content of calcites precipitated in the ocean. A major conclusion of their study is that under open ocean conditions the distribution of Mg-calcite cement is temperature controlled and that Mg-calcites with more than 10 mol% $MgCO_3$ tend to occur in waters less than 1000 m deep, that is above the base of the main thermocline. The only exceptions are semi-enclosed basins with warm bottom waters where Mg-calcite and aragonite are common.

Interpreting the age and depth of cemented limestones in the wall (deep fore-reef) and fore-reef (island slope) off Jamaica, Land & Moore (1977) have concluded that the sediments were lithified in pre-Holocene time. They suggest as a reason for this that decreased temperature and the poorly mixed state of water in the thermocline layer are not as conducive to the precipitation of submarine cement as waters in the shallow mixed layer above. The base of the shallow mixed layer off Jamaica lies at about 100 m and during the Pleistocene this would have been much lower than at present.

This conclusion would not seem to fit the data from Belize, where the base of the mixed surface layer is shallower, at a depth of about 50–70 m. Our data agree with the results from Jamaica in that much of the lithification occurred between 7000 and 15 000 years B.P. There is clear evidence, however, that lithification is still occurring on and in the wall; a partially lithified *Halimeda* grainstone, the last stage of a multi-generation cavity filling from the wall at a depth of 110·0 m (72–24·48) has a radiocarbon age of 2235 years B.P.; a *Montastraea annularis* colony at the surface of the wall, cemented hard to the wall rock at a depth of 97·5 m (75–207·10) has a radiocarbon age of 2430 years B.P. Both of these occurrences are well below the base of the mixed layer zone, even 2500 years B.P. In addition, the multigeneration process of boring, internal sedimentation and lithification should have ceased long ago on the wall if cementation was restricted to depths and palaeodepths shallower than 50 m.

The period between 7000 and 15 000 B.P. was a time when reefs were growing on the wall and on the fore-reef and so reef growth and sediment production in these areas was high. We believe that accretion was markedly reduced when maximum shallow to intermediate-water reef growth shifted slowly landward from the top of the wall as sea level rose. Sediments on the gentle fore-reef are not being cemented because sediments on the slope are moving slowly basinward, are being intensively bioturbated and have no fine-grained internal sediment matrix and the slope is, taken as a whole, not a site of slow or non-sedimentation.

In summary, this study has shown that in addition to being facies and fabric controlled early submarine cementation on the Belize carbonate platform is: (1) restricted to those parts of the margin above the base of the thermocline (generally less than 1000 m); (2) best developed in sediments that are deposited internally; (3)

enhanced if the deposition of these internal sediments is episodic; (4) brought about by the precipitation, not only of Mg-calcite, but of aragonite as well.

In the final analysis, from the viewpoint of petrography and our experience in phanerozoic limestones, this structure is already an ancient reef, yet it is only Holocene in age; most of the complex series of textures and structures so common to phanerozoic reefs are already present, yet the reef is still in the same diagenetic environment in which it grew.

REFERENCES

ADEY, W., MACINTYRE, I.G. & STUCKENRATH, R. (1977) Relict barrier reef system off St Croix: Its implications with respect to late Cenozoic coral reef development in the western Atlantic. In: *Proceedings of the Third International Coral Reef Symposium*, **2**, *Geology* (Ed. by D. L. Taylor), pp. 15–23. University of Miami, Miami.

AMIEL, A., FRIEDMAN, G.M. & MILLER, D.S. (1973) Distribution and nature of incorporation of trace elements in modern aragonite corals. *Sedimentology*, **20**, 47–65.

BAIE, L.F. (1970) Possible structural link between Yucatan and Cuba: *Bull. Am. Ass. Petrol. Geol.* **54**, 2204–2207.

BAK, R.P.M. (1977) Coral reefs and their zonation in the Netherlands Antilles. *Am. Ass. Petrol. Geol. Studies in Geology*, No. 4, 3–16.

BALL, M.M. (1967) Carbonate sand bodies of Florida and the Bahamas. *J. sedim. Petrol.* **37**, 556–591.

BARNES, J., BELLAMY, D.J., JONES, D.J., WHITTON, B.A., DREW, E.A., KENYON, L., LYTHGOE, J.N. & ROSEN, B.R. (1971) Morphology and ecology of the reef front of Aldabra. In: *Regional Variation in Indian Ocean Coral Reefs* (Ed. by D. R. Stoddart and C. M. Younge). *Symp. Zool. Soc. London*, **28**, 87–114.

BATTISTINI, R., BOURROUILH, F., CHEVALIER, J.P., COUDRAY, J., DENIZOT, M., FAURE, G., FISHER J.C., GUILCHER, A., HARMELIN-VIVIEN, M., JAUBERT, J., LABOREL, J., MASSE, J.P., MAUGE, L.A., MONTAGGIONI, L., PEYROT-CLAUSADE, M., PICHON, M., PLANTE, R., PLAZIAT, J.C., PLESSIS, Y.B., RICHARD, G., SALVAT, B., TOMASSIN, B.A., VASSEUR, P. & WEYDERT, P. (1975) *Terminologie Recifale Indopacifique, Tethys*, **7**.

BLOOM, A.L., BROECKER, W.S., CHAPPELL, J.M.A., MATTHEWS, R.K. & MESOLELLA, K.J. (1974) Quaternary sea level fluctuations on a tectonic coast: new $^{230}Th/^{234}U$ dates from the Huon Peninsula, New Guinea. *Quat. Res.* **4**, 185–205.

BOWIN, C.O. (1968) Geophysical study of the Cayman Trough. *J. geophys. Res.* **73**, 5159–5173.

BRADY, M.J. (1974) Sedimentation and diagenesis in slope and basinal environments off British Honduras *Abstr. Progr. geol. Soc. Am.* **5**, 97.

BUSBY, R.F. (1962) Submarine geology of the Tongue of the Ocean, Bahamas: *U.S. Navy Oceanog. Office Tech. Rep.* TR-108.

BUSBY, R.F. (1966) Ocean bottom reconnaissance off the east coast of Andros Island, Bahamas. *U.S. Navy Oceanog. Office Tech. Rep.* 20390.

CHAVE, K.E., SANDERS, H.L., HESSLER, R.R. & NEUMANN, A.C. (1962) Animal-sediment interrelationships on the Bermuda Slope and adjacent deep sea. *Bermuda Biological Station, ONR Report*, 10–12.

CHOQUETTE, P.W. & TRUSSELL, F.C. (1978) A procedure for making the Titan Yellow stain for Mg-calcite permanent. *J. sedim. Petrol.* **48**, 639–642.

CLOUD, P.E. (1952) Facies relationships of organic reefs. *Bull. Am. Ass. Petrol. Geol.* **36**, 2125–2149.

CURRAY, J.R. (1965) Late Quaternary history, continental shelves of the United States. In: *The Quaternary of the United States* (Ed. by H. E. Wright and R. Frey), pp. 723–735. Princeton University Press, Princeton, N.J.

CURRAY, J.R. & SHEPARD, F.P. (1972) Some major problems of Holocene sea levels. *Am. Quat. Ass. Second natn. Conf., Abstr.* 16–18.

D'ANGLEJAN, B.F. & MOUNTJOY, E.W. (1973) Submerged reefs of the Eastern Grenadines Shelf margin: *Bull. geol. Soc. Am.* **84**, 2445–2454.

DALY, P.A. (1910) Pleistocene glaciation and the coral reef problem. *Am. J. Sci.* **30**, 297–310.

DARWIN, C. (1842) *The Structure and Distribution of Coral Reefs*. Smith Elder and Co., London.

DAVID, T.W.E., HALLIGAN, G.H. & FINCKH, A.E. (1904) Report on dredging at Funafuti. In: *The Atoll of Funafuti*, Sect. VII, Royal Society of London, 151–159.

DAVIES, G.R. (1977) Turbidites, debris sheets and truncation structures in Upper-Paleozoic Deep-water Carbonates of the Sverdrup Arctic Archipelago. In: *Deep-water Carbonate Environments* (Ed. by H. E. Cook and P. Enos). *Spec. Publ. Soc. econ. Paleont. Miner., Tulsa*, **25**, 221–249.

DENGO, G. (1969) Problems of tectonic relations between Central America and the Caribbean: *Trans. Gulf-Cst Ass. geol. Socs, 19th Annual Mtg*, 311–320.

Dengo, G. & Bohnemberger, O. (1969) Structural development of northern Central America: *Mem. Am. Ass. Petrol. Geol.* **11,** 203–220.

Dill, R.F. (1971a) The Blue Hole: a structurally significant sink hole in the atoll off British Honduras: *Abstr. Progr. geol. Soc. Am.* **3,** 544–545.

Dill, R.F. (1971b) 'Blue Holes' Structurally significant solution features in calcareous platforms: *Abstr. Progr. geol. Soc. Am.* **3,** 545.

Dillon, W.P. & Vedder, J.G. (1973) Structure and development of the continental margin of British Honduras. *Bull. geol. Soc. Am.* **84,** 2713–2732.

Dreimanis, A. & Karrow, P.F. (1972) Glacial history of the Great Lakes–St Lawrence region, the classification of the Wisconsin (an) stage and its correlatives. *Int. Geol. Congr. 24th Session,* Sect. 12, 5–15.

Dunham, R.J. (1962) Classification of carbonate rocks according to depositional texture. In: *Classification of Carbonate Rocks.* (Ed. by W. H. Ham). *Mem. Am. Ass. Petrol. Geol.* **1,** 108–121.

Embry, A.F. & Klovan, J.E. (1971) A Late Devonian reef tract on northeastern Banks Island, N.W.T. *Bull. Can. Petrol. Geol.* **19,** 730–781.

Emery, K.O., Tracey, J.I. & Ladd, H.S. (1954) Geology of Bikini and nearby atolls: *Prof. Pap. U.S. geol. Surv.* **260-A.**

Emiliani, C. (1965) Precipitous continental slopes and considerations on the transitional crust. *Science,* **147,** 145–148.

Emiliani, C. (1951) On the species *Homotrema rubrum* (Lamarck). *Contribution of the Cushman Foundation for Foraminiferal Research,* **2,** 143–147.

Evamy, B.D. (1973) The precipitation of aragonite and its alteration to calcite on the Trucial Coast of the Persian Gulf. In: *The Persian Gulf* (Ed. by B. H. Purser), pp. 329–343. Springer-Verlag, Berlin.

Fairbridge, R.W. (1960) The changing level of the sea. Scientific American, **202,** 70–79.

Focke, J.W. (1978) Subsea (0–40 m) terraces and benches, windward off Curacao, Netherlands Antilles: *Leidse Geol. Mededelingen,* **51,** 95–102.

Folk, R.L. (1962) Spectral subdivision of limestone types. In: *Classification of Carbonate Rocks* (Ed. by W. E. Ham). *Mem. Am. Ass. Petrol. Geol.* **1,** 62–84.

Fox, P.J. & Schreiber, E. (1970) Granodiorites from the Cayman Trench. *Abstr. Progr. geol. Soc. Am.* **2,** 553.

Freeland, G.L. & Dietz, R.S. (1971) Plate tectonic evolution of Caribbean–Gulf of Mexico region. *Nature,* **232,** 20–23.

Friedman, G.M., Schneidermann, N. & Gervitz, J.L. (1971) Indurated hard calcium carbonate layers at the bottom of the Red Sea. In: *Carbonate Cements* (Ed. by O. P. Bricker), pp. 116–121. The Johns Hopkins University Press, Baltimore, Md.

Friedman, G.M., Amiel, A.J. & Schneidermann, N. (1974) Submarine cementation in reefs: example from the Red Sea: *J. sedim. Petrol.* **44,** 816–825.

Futterer, D.K. (1974) Significance of the boring sponge *Cliona* for the origin of fine-grained material of carbonate sediments. *J. sedim. Petrol.* **44,** 79–84.

Gardiner, J.S. (1903) The Maldive and Laccadive groups, with notes on the other coral formations in the Indian Ocean. In: *The Fauna and Geography of the Maldive and Laccadive Archipelagoes*: being an account of the work carried out on and of the collections made by an expedition during the years 1899 and 1900 (Ed. by J. S. Gardiner) pp. 12–50, 146–183, 313–346, 376–423. Cambridge University Press.

Gardiner, J.S. (1931) *Coral Reefs and Atolls.* Macmillan and Co., London.

Gavish, E. & Friedman, G.M. (1969) Progressive diagenesis in Quaternary to Late Tertiary carbonate sediments: sequence and time scale: *J. sedim. Petrol.* **39,** 49–69.

Gervitz, J.L. & Friedman, G.M. (1966) Deep sea carbonate sediments of the Red Sea and their importance on marine lithification. *J. sedim. Petrol.* **36,** 143–152.

Gibson, T.G. & Schlee, J. (1967) Sediments and fossiliferous rocks from the eastern side of the Tongue of the Ocean, Bahamas. *Deep Sea Res.* **14,** 691–702.

Ginsburg, R.N. (1953) Beachrock in south Florida. *J. sedim. Petrol.* **23,** 85–92.

Ginsburg, R.N. & James, N.P. (1973) British Honduras by submarine. *Geotimes,* **18,** 23–24.

Ginsburg, R.N. & James, N.P. (1974) Holocene carbonate sediments of continental shelves. In: *Continental Margins* (Ed. by C. A. Burk and C. L. Drake), pp. 137–154. Springer-Verlag, Berlin.

GINSBURG, R.N. & JAMES, N.P. (1976) Submarine botryoidal aragonite in Holocene reef limestones, Belize, *Geology,* **4,** 431–436.

GINSBURG, R.N. & SCHROEDER, J.H. (1973) Growth and submarine fossilization of algal cup reefs, Bermuda. *Sedimentology,* **20,** 575–614.

GINSBURG, R.N., SCHROEDER, J.H. & SHINN, E.A. (1971) Recent synsedimentary cementation in subtidal Bermuda reefs. In: *Carbonate Cements.* (Ed. by O. P. Bricker), pp. 54–59. The Johns Hopkins University Press, Baltimore, Md.

GINSBURG, R.N., SHINN, E.A. & SCHROEDER, J.H. (1968) Submarine cementation and intertidal cementation within Bermuda Reefs. (Abstr.) *Spec. Pap. geol. Soc. Am.* **115,** 78–79.

GOREAU, T.F. (1959) The ecology of Jamaican coral reefs. I. Species composition and zonation. *Ecology,* **40,** 67–90.

GOREAU, T.F. (1963) Calcium carbonate deposition by coralline algae and corals in relation to their role as reef builders. *N.Y. Acad. Sci.* **109,** 127–167.

GOREAU, T.F. & HARTMAN, W.D. (1963) Boring sponges as controlling factors in the formation and maintenance of coral reefs. In: *Maintenance of Hard Tissue Destruction,* pp. 25–54. *Am. Ass. Adv. Sci. Publs.* Washington D.C.

GOREAU, T.F. & HARTMAN, W.D. (1966) Sponge: effect on the form of reef corals. *Science,* **151,** 343–344.

GOREAU, T.F. & GRAHAM, E.A. (1967) A new species of *Halimeda* from Jamaica. *Bull. Mar. Sci.* **17,** 422–453.

GOREAU, T.F. & GOREAU, N.I. (1973) The ecology of Jamaican coral reefs. II. Geomorphology, zonation and sedimentary phases. *Bull. Mar. Sci.* **23,** 399–464.

GOREAU, T.F. & LAND, L.S. (1974) Fore-reef morphology and depositional processes, North Jamaica. In: *Reefs in Time and Space* (Ed. by L. F. Laporte). *Spec. Publ. Soc. econ. Paleont. Miner., Tulsa,* **18,** 77–89.

GRABAU, A.W. (1913) *Principles of Stratigraphy,* pp. 399–401. Dover Publications, New York (1960).

GROSS, M.G. (1964) Variations in the O^{18}/O^{16} and C^{13}/C^{12} ratios of diagenetically altered limestones in the Bermuda Islands. *J. Geol.* **72,** 170–194.

HALLEY, R.B., SHINN, E.A., HUDSON, J.H. & LIDZ, B. (1977) Recent and relict topography of Boo Bee patch reef, Belize. In: *Proceedings of the Third International Coral Reef Symposium,* **2,** *Geology* (Ed. by D. L. Taylor), pp. 29–35. University of Miami Press, Miami.

HENSON, F.R.S. (1950) Cretaceous and Tertiary reef formations and associated sediments in Middle East. *Bull Am. Ass. Petrol. Geol.* **34,** 215–238.

HESS, H.H. (1933) The submerged river valleys of the Bahamas. *Trans Am. geophs, Union 14th Ann. Mtg,* 168–170.

HOFFMAN, P. (1973) Recent and ancient algal stromatolites: seventy years of pedagogic cross-pollination. In: *Evolving Concepts in Sedimentology* (Ed. by R. N. Ginsburg), pp. 178–191. The Johns Hopkins Press, Baltimore, Md.

JAMES, N.P., GINSBURG, R.N., MARSZALEK, D.S. & CHOQUETTE, P.W. (1976) Facies and fabric specificity of early subsea cements in shallow Belize (British Honduras) reefs. *J. sedim. Petrol.* **46,** 523–544.

JAMES, N.P., STEARN, C.S. & HARRISON, R.S. (1977) *Field Guidebook to Modern and Pleistocene Reef Carbonates, Barbados, W.I., Third Int. Coral Reef Symp.,* University of Miami, Fisher Island, Miami.

KEITH, M.L. & WEBER, J.N. (1968) Carbon and oxygen isotopic composition of selected limestones and fossils. *Geochim. cosmochim. Acta,* **28,** 1787–1816.

KIER, J.S. & PILKEY, O.H. (1971) The influence of sea level changes on sediment carbonate mineralogy, Tongue of the Ocean, Bahamas. *Mar. Geol.* **11,** 189–200.

KINSMAN, D.J.J. (1969) Interpretation of Sr^{2+} concentrations in carbonate minerals and rocks. *J. sedim. Petrol.* **39,** 486–508.

KLEIN, G. deV (1971) A sedimentary model for determination of paleotidal range. *Bull. geol. Soc. Am.* **82,** 2582–2992.

KREBS, W. (1974) Devonian carbonate complexes of central Europe. In: *Reefs in Time and Space* (Ed. by L. F. Laporte). *Spec. Publ. Soc. econ. Paleont. Miner. Tulsa,* **18,** 155–209.

KREBS, W. & MOUNTJOY, E.W. (1972) Comparison of Central European and Western Canadian Devonian Reef Complexes. *24th Int. Geol. Congr.* Sect. 6, 294–309.

Ku, T.-L., & Broecker, W.S. (1969) Radiochemical studies on manganese nodules of deep sea origin. *Deep Sea Res.* **16,** 625–637.

Land, L.S. (1979) The fate of reef-derived sediment on the Jamaican Island Slope. *Mar. Geol.* **29,** 55–71.

Land, L.S. & Goreau, T.F. (1970) Submarine lithification of Jamaican reefs. *J. sedim. Petrol.* **40,** 457–462.

Land, L.S. & Hoops, G.K. (1973) Sodium in carbonate sediments and rocks: a possible index to the salinity of diagenetic solutions. *J. sedim. Petrol.* **43,** 614–617.

Land, L.S. & Moore, C.H. (1974) Modern Jamaican biolithites and deep reef talus deposits: petrography and geochemistry. *Ann. Mtg. Abstr. Am. Ass. Petrol. Geol.* **56.**

Land, L.S. & Moore, C.H. (1977) Deep Fore-reef and Upper Island Slope, North Jamaica. In: *Reefs and Related Carbonates: Ecology and Sedimentology* (Ed. by S. H. Frost, M. P. Weiss and J. B. Saunders). *Am. Ass. Petrol. Geol. Studies in Geology* No. 4, 53–67.

Lang, J. (1974) Biological zonation at the base of a reef. *Am. Scient.* **63,** 272–281.

Lighty, R.G. (1977) Relict shelf-edge Holocene coral reef, southeast coast of Florida. In: *Proceedings of the Third International Coral Reef Symposium,* **2,** *Geology* (Ed. by D. L. Taylor), pp. 215–223. University of Miami Press, Miami.

Link, T.A. (1950) Theory of transgressive and regressive reef (Bioherm) development and origin of oil. *Bull. Am. Ass. Petrol. Geol.* **34,** 263–294.

Logan, B.W. (1969) Coral reefs and banks, Yucatan Shelf, Mexico. In: *Carbonate Sediments and Reefs, Yucatan Shelf, Mexico* (Ed. by B. W. Logan, J. L. Harding, W. M. Alan, J. D. Williams & R. G. Snead). *Mem. Am. Ass. Petrol. Geol.* **11,** 129–196.

Logan, B.W., Rezak, R. & Ginsburg, R.N., (1964) Classification and environmental significance of algal stromatolites. *J. Geol.* **72,** 68–83.

Lowenstam, H.A. & Epstein, S. (1957) On the origin of sedimentary aragonite needles of the Great Bahama Bank. *J. Geol.* **65,** 364–375.

Lucia, F.J. (1972) Recognition of evaporite-carbonate shoreline sedimentation. In: *Recognition of Ancient Sedimentary Environments* (Ed. by J. K. Rigby and W. K. Hamblin). *Spec. Publ. Soc. econ. Paleont. Miner., Tulsa,* **16,** 160–191.

Macintyre, I.G. (1967) Submerged coral reefs, west coast of Barbados, West Indies. *Can. J. Earth Sci.* **4,** 461–474.

Macintyre, I.G. (1972) Submerged reefs of eastern Caribbean. *Bull. Am. Ass. Petrol. Geol.* **56,** 720–738.

Macintyre, I.G. (1977) Distribution of submarine cements in a modern Caribbean fringing reef, Galeta Point, Panama. *J. sedim. Petrol.* **47,** 503–516.

Macintyre, I.G., Mountjoy, E.W. & D'Anglejan, B.F. (1968) An occurrence of submarine cementation of carbonate sediments off the west coast of Barbados, W.I. *J. sedim. Petrol.* **38,** 660–663.

Mackenzie, F.T., Kulm, L.D., Cooley, R.L. & Barnhart, J.T. (1965) *Homotrema rubrum:* a sediment transport indicator: *J. sedim Petrol.* **35,** 265–272.

Marlowe, J.I. (1971) High-magnesian calcite cement in calcarenites from Aves Swell, Caribbean Sea. In: *Carbonate Cements* (Ed. by O. P. Bricker), pp. 111–116. The Johns Hopkins University Press, Baltimore, Md.

Matthews, R.K. (1965) Genesis of Recent lime mud in southern British Honduras. *J. sedim. Petrol.* **36,** 428–454.

Maxwell, W.G.H. (1968) *Atlas of the Great Barrier Reef.* Elsevier, Publishing Co., New York.

Meischner, D. & Meischner, U. (1977) Bermuda South Shore Reef Morphology. In: *Proceedings of the Third International Coral Reef Symposium* **2,** *Geology* (Ed. by D. L. Taylor). pp. 243–250. University of Miami Press, Miami.

Mesolella, K.J., Sealy, H.A. & Matthews, R.K. (1970) Facies geometries within Pleistocene Reefs of Barbados, West Indies. *Bull. Am. Ass. Petrol. Geol.* **54,** 1899–1918.

Miller, J.A. & Macintyre, I.G. (1977) *Field Guidebook to the Reefs of Belize. Third Int. Coral Reef Symp.* University of Miami, Fisher Island, Miami.

Milliman, J.D. (1966) Submarine lithification of carbonate sediments. *Science,* **153,** 994–997.

Milliman, J.D. (1974) *Marine Carbonates.* Springer-Verlag, Berlin.

Milliman, J.D. & Emery, K.O. (1968) Sea levels during the past 35 000 years. *Science,* **162,** 1121–1123.

MOLNAR, P. & SYKES, L.R. (1969) Tectonics of the Caribbean and Middle America regions from local mechanisms and seismicity. *Bull. geol. Soc. Am.* **80,** 1639–1684.

MOORE, C.H., GRAHAM, E.A. & LAND, L.S. (1976) Sediment transport and dispersal across the deep fore-reef and island slope (−55 m to 305 m), Discovery Bay, Jamaica, *J. sedim. Petrol.* **46,** 174–187.

MORNER, N. (1971) Eustatic changes during the last 20 000 years and a method of separating the isostatic and eustatic factors in an uplifted area. *Palaeogeogr. Palaeoclim. Palaeoecol.,* **9,** 153–181.

MORNER, N. (1976) Eustasy and geoid changes. *J. Geol.* **84,** 123–151.

MULLINS, H.T. & NEUMANN, A.C. (1979), Carbonate slopes along open seas and seaways in the Northern Bahamas. *Spec. Publ. Soc. econ. Paleont. Miner., Tulsa.* (In press.)

NEUMANN, A.C. (1965) Processes of Recent carbonate sedimentation in Harrington Sound, Bermuda. *Bull. Mar. Sci.* **15,** 987–1035.

NEUMANN, A.C. & BALL, M.M. (1970) Submersible observations in the Straits of Florida: geology and bottom currents. *Bull. geol. Soc. Am.* **81,** 2861–2874.

NEWELL, N.D., RIGBY, J.K., WHITEMAN, A.J. & BRADLEY, J.S. (1951) Shoal water geology and environments, eastern Andros Island, Bahamas. *Bull. Am. Mus. Nat. Hist.* **97,** 1–29.

NEWELL, N.D. & RIGBY, J.K. (1957) Geological studies on the Great Bahama Bank. In: *Regional Aspects of Carbonate Deposition* (Ed. by R. J. Leblanc and J. G. Breeding). *Spec. Publ. Soc. econ. Paleont. Miner., Tulsa,* **5,** 15–75.

PLAYFORD, P.E. (1969) Devonian carbonate complexes of Alberta and Western Australia: a comparative study. *Geol. Surv. W. Aust. Rep.* **1.**

PRAY, L.C. & MURRAY, R.C. (1965) *Dolomitization and Limestone Diagenesis:* a Symposium. *Spec. Publ. Soc. econ. Paleont Miner., Tulsa,* **13.**

PURDY, E.G. (1974a) Reef configurations: cause and effect: *Spec. Publ. Soc. econ. Paleont. Miner., Tulsa,* **18,** 9–76.

PURDY, E.G. (1974b) Karst determined facies patterns in British Honduras: Holocene carbonate sedimentation model. *Bull. Am. Ass. Petrol. Geol.* **58,** 825–855.

PURDY, E.G., PUSEY, W.C. & WANTLAND, K.F. (1975) Continental shelf of Belize: regional shelf attributes. In: *Belize Shelf: Carbonate Sediments, Clastic Sediments and Ecology* (Ed. by K. F. Wantland and W. C. Pusey). *Am. Ass. Petrol. Geol. Studies in Geology,* No. 2, 1–40.

PUSEY, W.C. (1964) *Recent calcium carbonate sedimentation in northern British Honduras.* Unpublished Ph.D. Thesis, Rice University, Houston.

PUSEY, W.C., EBANKS, W.J., HIGH, L. & SCOTT, M. (1975) Holocene carbonate sedimentation on northern Belize Shelf. In: *Belize Shelf: Carbonate Sediments, Clastic Sediments, and Ecology* (Ed. by K. F. Wantland and W. C. Pusey). *Am. Ass. Petrol. Geol. Studies in Geology,* No. 2, 131–234.

REUTZLER, K. & RIEGER, G. (1973) Sponge burrowing: fine structure of *Cliona lampa* penetrating calcareous substrata. *Mar. Biol.* **21,** 144–162.

RIGBY, J.K. & ROBERTS, H.H. (1976) *Grand Cayman Island: Geology, Sediments and Marine Communities: Spec. Publ. Brigham Young University Geology Studies,* No. 4.

ROONEY, W.S. (1970) A preliminary ecologic and environmental study of the sessile foraminifer, *Homotrema rubrum* (Lamarck). In: *Seminar on Organism-Sediment Interrelationships* (Ed. by R. N. Ginsburg and S. M. Stanley) *Spec. Publ. Bermuda Biol. Station* No. 6, 7–19.

ROSEN, B.R. (1971) The distribution of reef coral genera in the Indian Ocean.: In: *Regional Variation in Indian Ocean Coral Reefs* (Ed. by D. R. Stoddart and C. M. Yonge). *Symp. Zool. Soc. London,* **28,** 263–299.

SCHLAGER, W., HOOKE, R.L. & JAMES, N.P. (1976) Episodic erosion and deposition in the Tongue of the Ocean (Bahamas). *Bull. geol. Soc. Am.* **87,** 1115–1118.

SCHLAGER, W. & JAMES, N.P. (1978) Low-magnesian calcite limestones forming at the deep-sea floor, Tongue of the Ocean, Bahamas. *Sedimentology,* **25,** 675–702.

SCHLAGER, W. & CHERMAK, A. (1979) Sediment facies of platform-basin transition, Tongue of the Ocean, Bahamas. *Mar. Geol.* (In press.)

SCHLAGER, W., ASSERETO, R., JAMES, N.P., LAND, J., SHINN, E.A. & SLATER, R. (in preparation) Morphology and composition of the upper marginal escarpment, Tongue of the Ocean, Bahamas.

SCHMIDT, V. (1971) Early carbonate cementation in Middle Devonian bioherms, Rainbow Lake, Alberta. In: *Carbonate Cements* (Ed. by O. P. Bricker), pp. 209–216. The Johns Hopkins University Press, Baltimore, Md.

SCHOLLE, P.A. & KLING, S.A. (1972) Southern British Honduras: lagoonal coccolith ooze. *J. sedim. Petrol.* **42,** 195–204.

SCHROEDER, J.H. (1972) Fabrics and sequences of submarine carbonate cements in Holocene Bermuda cup reefs. *Geol. Rdsch.* **61,** 708–730.

SHINN, E.A. (1969) Submarine lithification of Holocene carbonate sediments in the Persian Gulf. *Sedimentology,* **12,** 109–144.

STANLEY, D.J. & SWIFT, D.J.P. (1968) Bermuda's reef-front platform: bathymetry and significance. *Mar. Geol.* **6,** 479–500.

STODDART, D.R. (1962a) Catastrophic storm effects on British Honduras reefs and cays. *Nature,* **196,** 512–515.

STODDART, D.R. (1962b) Three Caribbean Atolls: Turneffe Islands, Lighthouse Reef and Glovers Reef, British Honduras. *Atoll Res. Bull.* No. 87, 1–147.

STODDART, D.R. (1963) Effects of Hurricane Hattie on the British Honduras reefs and cays. October 30–31, 1961. *Atoll Res. Bull.* **95,** 1–142.

STODDART, D.R. (1969) Ecology and morphology of recent coral reefs. *Biol. Rev.* **44,** 433–498.

STODDART, D.R. & CANN, J.R. (1965) Nature and origin of beachrock. *J. sedim. Petrol.* **35,** 233–247.

SYKES, L.R. & EWING, M. (1965) The seismicity of the Caribbean region. *J. geophys. Res.* **70,** 5065–5074.

TAYLOR, J.C.M. & ILLING, L.V. (1971) Alteration of recent aragonite to magnesian calcite cement, Qatar, Persian Gulf. In: *Carbonate Cements* (Ed. by O. P. Bricker), pp. 36–40. The Johns Hopkins University Press, Baltimore, Md.

UCHUPI, E. (1973) Eastern Yucatan continental margin and western Caribbean tectonics. *Bull. Am. Ass. Petrol. Geol.* **57,** 1075–1085.

VAN ANDEL, T.H. & VEEVERS, J.J. (1967) Morphology and sediments of the Timor Sea. *Bur. Miner. Res. W. Aust. Geol. Geophys. Bull.* **83,** 1–173.

VAUGHAN, T.W. (1907) Some madreoporian corals from French Somaliland, East Africa, collected by Dr Charles Gravier. *Proc. U.S. natn. Mus.* **32,** 249–266.

VAUGHAN, T.W. (1919) Fossil corals from Central America and Puerto Rico, with an account of the American Tertiary, Pleistocene and Recent coral reefs. *Bull. U.S. natn. Mus.* **103,** 189–524.

VEDDER, J.G., LANPHERE, M.A., MACLEOD, N.S. & DILLON, W.P. (1973) Age and tectonic implications of some low-grade metamorphic rocks from the Yucatan Channel. *J. Res. U.S. geol. Surv.* **1,** 157–164.

VINIEGRA, F. (1971) Age and evolution of salt basins of southeastern Mexico. *Bull. Am. Ass. Petrol. Geol.* **55,** 478–494.

WALCOTT, R.I. (1972) Past sea levels and deformation of the earth. *Quat. Res.* **2,** 1–14.

WALLACE, R.J. & SCHAFERSMANN, S.D. (1977) Patch-reef ecology and sedimentology of Glovers Reef Atoll, Belize. In: *Reefs and Related Carbonates: Ecology and Sedimentology* (Ed. by S. H. Frost, M. P. Weiss and J. B. Saunders). *Am. Ass. Petrol. Geol. Studies in Geology* No. 4, 37–53.

WANTLAND, K.F. & PUSEY, W.C. III (1971) A guidebook for the field to the southern shelf of British Honduras. *Gulf-Cst Ass. geol. Socs.* 21*st Ann. Mtg,* 1–87.

WARBURTON, F.E. (1958) The manner in which *Cliona* bores into objects. *Can. J. Zool.* **36,** 555–562.

WEBER, J.N. (1967) Factors affecting the carbon and oxygen isotopic composition of marine carbonate sediments. I. Bermuda. *Am. J. Sci.* **265,** 586–608.

WEBER, J.N. & WOODHEAD, P.M.J. (1969) Factors affecting the carbon and oxygen isotopic composition of marine carbonate sediments. II. Heron Island, Great Barrier Reef, Australia. *Geochim. cosmochim. Acta,* **33,** 19–38.

WEBER, J.N. & WOODHEAD, P.M.J. (1970) Carbon and oxygen isotope fractionation in the skeletal carbonate of reef-building corals. *Chem. Geol.* **6,** 93–117.

WELLS, J.W. (1954) Bikini and nearby atolls: (2) Oceanography (biologic), Recent corals of the Marshall Islands. *Prof. Pap. U.S. geol. Surv.* **260-I,** 385–486.

WELLS, J.W. (1957) Coral reefs. In: *Treatise on Marine Ecology and Paleoecology* (Ed. by J. W. Hedgepeth). *Mem. geol. Soc. Am.* **67,** 609–633.

WILSON, J.L. (1969) Microfacies and sedimentary structures in 'Deeper water' lime mudstones. In: *Depositional Environments in Carbonate Rocks* (Ed. by G. M. Friedman). *Spec. Publ. Soc. econ. Paleont. Miner., Tulsa,* **14,** 4–17.

WINLAND, H.D. (1971) Nonskeletal deposition of high Mg-calcite in the marine environment and its role in the retention of textures. In: *Carbonate Cements* (Ed. by O. P. Bricker), pp. 278–285. The Johns Hopkins University Press, Baltimore, Md.

WRAY, J.L., JAMES, N.P. & GINSBURG, R.N. (1975) The puzzling Paleozoic phylloid algae: an answer in Holocene Squamariacean calcareous red algae. *Am. Ass. Petrol. Geol. Ann. Mtg. Abstr.* **2,** 82–83.

WRAY, J.L. (1977) Late Paleozoic calcareous red algae. In: *Fossil Algae* (Ed. by E. Flugel, pp. 167–176. Springer-Verlag, Berlin.

WRIGHT, A.C.S., ROMNEY, D.H., ARBUCKLE, R.H. & VIAL, V.E. (1959) Land in British Honduras. *Colonial Research Publs* **24,** Colonial Office, London.

WUST, G. (1964) *Stratification And Circulation In The Antillean–Caribbean Basins.* Columbia University Press, New York.

ZANKL, H. (1969) Der Hohe Göll Aufbau und Lebensbild eines Dachstein-Riffes in der Obertrias der Nördlichen Kalkalpen-Abh. *Senckenberg. Naturforsch. Ges.* **519,** 1–123.

ZANKL, H. (1971) A model of sedimentation and diagenesis in a Triassic reef. In: *Carbonate Cements* (Ed. by O. P. Bricker), pp. 189–193. The Johns Hopkins University Press, Baltimore, Md.

ZANKL, H. & SCHROEDER, J.H. (1972) Interaction of genetic processes in Holocene reefs off North Eleuthera Island Bahamas. *Geol. Rsch.* **61,** 520–541.